Nonlinear Instabilities in Plasmas and Hydrodynamics

Plasma Physics Series

Series Editors:

Professor Peter Stott, CEA Cadarache, France
Professor Hans Wilhelmsson, Chalmers University of Technology, Sweden

Other books in the series

An Introduction to Alfvén Waves
R Cross

Transport and Structural Formation in Plasmas
K Itoh, S-I Itoh and A Fukuyama

Tokamak Plasma: a Complex Physical System
B B Kadomtsev

Electromagnetic Instabilities in Inhomogeneous Plasma
A B Mikhailovskii

Instabilities in a Confined Plasma
A B Mikhailovskii

Physics of Intense Beams in Plasma
M V Nezlin

Forthcoming titles in the series in 2000

Plasma Physics via Computer Simulation, 2nd Edition
C K Birdsall and A B Langdon

Particle Transport in Magnetically Confined Plasmas
G L Fussmann

Laser-Aided Diagnostics of Gases and Plasmas
K Muraoka and M Maeda

Inertial Confinement Fusion
S Pfalzner

Introduction to Dusty Plasma Physics
P K Shukla and N Rao

The Plasma Boundary of Magnetic Fusion Devices
P Stangeby

Collective Modes in Inhomogeneous Plasma: Kinetic and Advanced Fluid Theory
J Weiland

Plasma Physics Series

Nonlinear Instabilities in Plasmas and Hydrodynamics

S S Moiseev

Space Research Institute, Moscow, Russia

V G Pungin

IZMIRAN, Troitsk, Moscow Region, Russia

Victor N Oraevsky

Space Research Institute, Moscow, Russia

Translated from the Russian by Vitaly I Kisin

CRC Press
Taylor & Francis Group
Boca Raton London New York

CRC Press is an imprint of the
Taylor & Francis Group, an informa business

British Library Cataloguing-in-Publication Data

A catalogue record for this book is available from the British Library.

ISBN 0 7503 0483 9

Library of Congress Cataloging-in-Publication Data are available

Reprinted 2010 by CRC Press
CRC Press
6000 Broken Sound Parkway, NW
Suite 300, Boca Raton, FL 33487
270 Madison Avenue
New York, NY 10016
2 Park Square, Milton Park
Abingdon, Oxon OX14 4RN, UK

Production Editor: Al Troyano
Production Control: Sarah Plenty and Jenny Troyano
Commissioning Editor: Michael Taylor
Editorial Assistant: Victoria Le Billon
Cover Design: Jeremy Stephens
Marketing Executive: Colin Fenton

Published by Institute of Physics Publishing, wholly owned by The Institute of Physics, London

Institute of Physics Publishing, Dirac House, Temple Back, Bristol BS1 6BE, UK

US Office: Institute of Physics Publishing, The Public Ledger Building, Suite 1035, 150 South Independence Mall West, Philadelphia, PA 19106, USA

Typeset in TeX using the IOP Bookmaker Macros

Contents

Preface

The clear understanding of the important role played by collective wave processes in physical phenomena in the solid, liquid, gaseous and plasma states of matter constituted one of the most important stages in the progress of physics in the second half of the 20th century. The concepts of oscillations and oscillatory systems were replaced by the concepts of waves and wave-sustaining media. The first to be studied were the properties of wave-sustaining media with small perturbations: the so-called linear properties of wave-sustaining media. However, the advancement of technologies made it possible to create high-power perturbations that produced important changes in the nature of physical processes in wave-sustaining media. First, high-power perturbations made a medium non-equilibrium. A natural consequence of the non-equilibrium character of a medium is the development of instabilities which lead to the fastest possible removal of non-equilibrium. Second, the medium becomes nonlinear, which, strictly speaking, signifies that the principle of superposition is no longer valid and that various channels of interaction are created. Interactions are generated not only between the individual degrees of freedom but also between the collective and individual degrees of freedom. This is the reason why a powerful perturbation can 'discharge' not only the non-equilibrium that this perturbation has generated (we will refer to it as the nonlinear non-equilibrium) but can also help 'discharge' the non-equilibrium that existed before the perturbation (let us call it the initial non-equilibrium). Correspondingly, we subsume nonlinear instabilities into two types: (1) *direct* nonlinear instabilities due to the 'discharge' of a nonlinear non-equilibrium, and (2) *induced* nonlinear instabilities[1] due to the non-equilibrium nature of the initial medium; these arise in the presence of an external perturbation or if there are at least two parameters of non-equilibrium.

Among instabilities of the *first* type we find the parametric decay instability, the modulational instability and the entire large class of parametric instabilities of high-amplitude waves (Oraevsky 1984).

Instabilities of the *second* type (Kogan *et al* 1965, Moiseev and Sagdeev 1986, Moiseev *et al* 1981)—induced nonlinear instabilities—arise in systems that were non-equilibrium from the start but where the initial instability has already

[1] Sometimes colloquially known as secondary instabilities.

reached the saturation regime. Nonlinear processes—typically high-frequency—occurring in such systems often generate high-frequency turbulence. One of the known consequences of turbulence is scattering of particles (e.g. electrons and ions in plasmas) by turbulent dynamic inhomogeneities and vortices. This leads to naturally arising turbulent dissipative processes described by characteristic times and their reciprocals—effective collision frequencies.

For example, if plasma displays only high-frequency turbulence that affects only collective wave motion of electrons, then electron scattering on such turbulent dynamic inhomogeneities produces momentum exchange within the electron component. One of the consequences is the generation of the so-called anomalous electron viscosity (Kogan *et al* 1963) in the absence of anomalous conduction (there is no friction between the electron and ion components). Kogan *et al* (1963) were able to show that drift-type instability is then generated in magnetoactive inhomogeneous plasmas. This was historically the first example of an induced nonlinear instability of hydrodynamic type. It was later shown that induced nonlinear instabilities arise not only in plasmas but in hydrodynamics as well, and play an important role in the dynamics of nonlinear wave-sustaining media.

Another, and quite spectacular, example of induced instabilities is the explosive instability of negative-energy waves. Negative-energy waves earned this name because, as their amplitude grows (at constant energy of the medium), the total energy of the medium + wave system decreases (Kadomtsev *et al* 1964). These waves arise in non-equilibrium systems (for details, see section 1.5). Negative-energy waves possess a number of peculiar features. For instance, taking dissipation into account leads not to damping but to self-excitation of these waves. Indeed, dissipation must increase the energy of the medium while the total energy of the medium + wave system remains constant. Therefore, an increase in the medium's energy must be compensated for by an increase in the negative energy of the wave. The explosive instability (Coppi *et al* 1969, Karpliuk *et al* 1970) arises when the energy of the negative-energy wave is 'channelled' as a result of three-wave interactions to ordinary waves, that is, the negative-energy wave is nonlinearly dissipated.

The aim of this book is to present in a systematic manner the basic ideas of the physics of nonlinear instabilities in wave-sustaining media. We would be very naive if we wished to unfold the vast data available in this field within a relatively small monograph. Our presentation pays most attention to the ideas connected with induced nonlinear instabilities. This choice was dictated by the fact that these aspects have not been adequately developed until now in reviews and books devoted to nonlinear wave-sustaining media. We thus strive to summarize, for the first time, the results available in papers scattered through journals, preprints and proceedings of various conferences.

It is only natural that whenever the ideas of the physics of nonlinear instabilities are presented, we always accompany this with specific calculations and give an adequately detailed description of the mathematics involved.

References

Coppi B, Rosenbluth M N and Sudan R N 1969 *Ann. Phys.* **55** 207

Kadomtsev B B, Mikhailovsky A B and Timofeev A V 1964 *Zh. Exp. Teor. Fiz.* **47** 2266–8

Karpliuk K S, Oraevsky V N and Pavlenko V P 1970 *Ukr. Fiz. Zh.* **15** 340

Kogan E Ya, Moiseev S S and Oraevsky V N 1965 *Zh. Prikl. Mech. Tech. Fiz.* **6** 41–6

Moiseev S S and Sagdeev R Z 1986 *Izv. Vuzov, Radiofiz.* **29** 1067–72

Moiseev S S, Sagdeev R Z, Tur A V and Yanovsky V V 1981 *Zh. Exp. Teor. Fiz.* **80** 597–607

Oraevsky V N 1984 *Handbook of Plasma Physics* vol 2, ed M N Rosenbluth and R Z Sagdeev (Amsterdam: North-Holland) pp 37–79

Chapter 1

Direct nonlinear instabilities

1.1 Parametric decay instability

The hierarchy of nonlinear instabilities can be traced back to the parametric decay instability discovered in 1962 (Oraevsky and Sagdeev 1962). The instability arises when the phase and frequency matching are produced for the initial wave, usually called the *pump wave*, and two other eigenmodes of the system whose amplitudes are initially at the level of thermal fluctuations. The conditions can be written as (Oraevsky and Sagdeev 1962)

$$\begin{aligned} \omega_0 &= \omega_1 + \omega_2 \\ \boldsymbol{k}_0 &= \boldsymbol{k}_1 + \boldsymbol{k}_2 \end{aligned} \tag{1.1.1}$$

where ω_0 and $\boldsymbol{k}_0$ are the frequency and wave vector of the pump wave, ω_1 and $\boldsymbol{k}_1$ and ω_2 and $\boldsymbol{k}_2$ are the corresponding quantities of the eigenmodes of the medium.

Conditions (1.1.1) coincide with the laws of conservation for energy and momentum which hold for decays of quasi-particles. For this reason, instabilities of this type are traditionally referred to, after the work of Oraevsky and Sagdeev (1962), as decay instabilities.

In the same year when the parametric decay instability (PDI) in plasma was discovered, the theory of stimulated scattering of waves was developed independently in nonlinear optics (Kroll 1962, Woodbary and Ng 1962). It soon became clear that the decay instability (DI) forms the basis of the processes of stimulated Raman wave scattering. Those features of the stimulated Raman wave scattering that were not quite understood before (since Einstein classified processes into spontaneous and stimulated) are in fact caused by the nature of wave processes under decay instability. It is precisely owing to decay instability that we observe the exponential (not linear!) growth of the amplitudes of both scattered and incident waves! This is the immediate consequence of the positive feedback of the scattered and incident waves that propagate against the background of the pump wave. The equations that describe this relationship are in fact the spatial-temporal generalization of the Hill equations. The corresponding instabilities are naturally classified as parametric. In the general case, parametric instabilities of this type form if

$$\begin{aligned} n\omega_0 &= \omega_1 + \omega_2 \\ n\boldsymbol{k}_0 &= \boldsymbol{k}_1 + \boldsymbol{k}_2 \end{aligned} \qquad n = 1, 2, 3 \ldots. \tag{1.1.2}$$

Instabilities of practical importance for moderate-amplitude pump waves are the first-order instabilities ($n = 1$), that is, decay instabilities. Artsimovich and Sagdeev (1979) proved that in the absence of decay instability, second-order instabilities arise.

1.2 Decay instabilities in media with wave-modulated phase velocity; qualitative theory and principal features of PDI

We know that the Mathieu equation describes parametric instabilities in oscillatory systems. It is usually written in the form

$$\ddot{x} + \omega^2[1 + \varepsilon\cos(\omega_0 t)]x = 0. \tag{1.2.1}$$

A natural (and the simplest) generalization of this equation to wave-sustaining media is the equation

$$\frac{\partial^2 u}{\partial t^2} - V_\Phi^2[1 + \varepsilon\cos(\omega_0 t - k_0 x)]\frac{\partial^2 u}{\partial x^2} + \hat{\alpha} u = 0 \tag{1.2.2}$$

where $\hat{\alpha} u$ is a linear operator describing the deviation of the wave dispersion from the linear law $\omega = kV_\Phi$. In the absence of a term proportional to ε, equations of the form of (1.2.2) describe linear properties of waves, for example, of acoustic waves in gas dynamics, or magnetoacoustic and Alfven waves in magnetic hydrodynamics. For instance, the one-dimensional approximation of the equations of gas dynamics (including the gas dynamics of isotropic plasma),

$$\left.\begin{aligned} &\frac{\partial \rho}{\partial t} + \rho_0\frac{\partial u}{\partial x} = 0 \\ &\rho_0\frac{\partial u}{\partial t} = -\frac{\partial p}{\partial x} \qquad \frac{P}{\rho^\gamma} = \text{const} \end{aligned}\right\} \tag{1.2.3}$$

where ρ, u and p are the density, mass velocity and pressure, respectively, γ is the exponent of the adiabatic curve and the subscript zero indicates undisturbed quantities, gives the equation

$$\frac{\partial^2 u}{\partial t^2} - V_\Phi^2\frac{\partial^2 u}{\partial x^2} = 0 \tag{1.2.4}$$

where

$$V_\Phi^2 = s^2 \equiv \gamma\frac{P_0}{\rho_0} \qquad (s \text{ is the speed of sound}).$$

Another example. Let us concentrate for a moment on low-amplitude Alfven waves. (Such movements *per se* are incompressible, div $\boldsymbol{V} = 0$.) We denote by $\boldsymbol{h}$ the magnetic field of a wave that propagates in a constant magnetic field $\boldsymbol{H}_0$. Using the equations

$$\begin{aligned}\frac{\partial \boldsymbol{u}}{\partial t} &= \left(\frac{1}{4\pi\rho_0}\right)[\operatorname{rot}\boldsymbol{h}, \boldsymbol{H}_0] \\ \frac{\partial \boldsymbol{h}}{\partial t} &= \operatorname{rot}[\boldsymbol{u}, \boldsymbol{H}_0]\end{aligned} \tag{1.2.5}$$

we obtain

$$\frac{\partial^2 u}{\partial t^2} - V_\Phi^2 \frac{\partial^2 u}{\partial x^2} = 0 \qquad V_\Phi^2 = V_A^2 = \frac{H_0^2}{4\pi\rho_0}. \tag{1.2.6}$$

Here the axis x points along $\boldsymbol{H}_0$ and u denotes any component of the mass velocity that is perpendicular to $\boldsymbol{H}_0$.

Assume now that we have modulated the density of the medium by a low-amplitude sine wave (pump wave). We have

$$\begin{aligned}s^2 &= s_0^2[1 + \varepsilon\cos(\omega_0 t - k_0 x)] \\ V_A^2 &= V_{A_0}^2[1 + \varepsilon\cos(\omega_0 t - k_0 x)] \\ \varepsilon &= -\frac{\delta\rho_0}{\rho_0}\end{aligned} \tag{1.2.7}$$

where $\delta\rho_0$ is the pump wave amplitude.

If we formally substitute (1.2.7) into (1.2.5) and (1.2.6), we arrive at an equation of type (1.2.2).

Of course, if the equations for acoustic or Alfven waves in media with wave modulation are rigorously deduced, then (as we demonstrate later) additional terms appear, involving nonlinearities that are not of the type considered above. However, this factor will not change the principal conclusions yielded by the qualitative analysis outlined in this section.

We thus begin with equation (1.2.2). We shall illustrate by this example how the parametric coupling arises in the wave doublet $(\omega_1, \boldsymbol{k}_1; \omega_2, \boldsymbol{k}_2)$ described by equation (1.2.2). We see that in the absence of pump waves, that is, if $\varepsilon = 0$, equation (1.2.2) describes plane waves obeying the dispersion law $\omega(k) = kV_\Phi + \alpha(k)$, where $\alpha(k)$ is the contribution to frequency due to the terms generated by the operator (see (1.2.2)).

To investigate the parametric coupling of waves due to the pump wave background, we change to the Fourier transforms in spatial variables, $V_k = \int u(x)\exp(ikx)\,dx$, and separate the term due to the pump wave, transferring it to the right-hand side of the equation. The result is

$$\begin{aligned}\frac{d^2 V_{k_1}}{dt^2} + \omega^2(k_1)V_{k_1} &= \left(\frac{\varepsilon}{2}\right)(k_0 - k_1)^2 V_{\Phi_0}^2 V_{k_0-k_1}^* e^{-i\omega_0 t} \\ &\quad - \left(\frac{\varepsilon}{2}\right)(k_0 + k_1)^2 V_{\Phi_0}^2 V_{k_0+k_1}^* e^{-i\omega_0 t}.\end{aligned} \tag{1.2.8}$$

The set (1.2.8) is the set of equations for coupled oscillators. Following Oraevsky, it can be truncated, in view of the smallness of the parameter ε. We see that for the zero approximation in ε, V_k oscillates in time at eigenfrequencies $\omega(k)$. In ordinary non-resonant conditions, weak coupling influences the wave dynamics only slightly. However, if the driving force in the right-hand side happens to be in resonance with an eigenfrequency, the oscillator may switch to the excitation mode. Obviously, the resonance condition for the first term is $\omega_0 - \omega(k_0 - k_1) = \omega(k_1)$; for the second term it is $\omega_0 - \omega(k_1 + k_0) = \omega(k_1)$. For certainty, we assume that the first condition holds. The second term is then non-resonant, and thus can be dropped.[1] Now we need to add to equation (1.2.8) for V_{k_1} an equation for $V^*_{k_0-k_1}$:

$$\frac{d^2 V^*_{k_0-k_1}}{dt^2} + \omega^2(k_0 - k_1)V^*_{k_0-k_1} = -\left(\frac{\varepsilon}{2}\right) k_1^2 V_{\Phi_0}^2 e^{i\omega_0 t} V_{k_1} - \left(\frac{\varepsilon}{2}\right)(2k_0 - k_1)^2 V_{\Phi_0}^2 e^{-i\omega_0 t} V_{2k_0-k_1}. \tag{1.2.9}$$

The second term in this equation is non-resonant as well. Therefore, keeping only resonant terms in (1.2.8) and (1.2.9), we arrive at the following truncated equation for $V^*_{k_0-k_1}$:

$$\begin{aligned} \frac{d^2 V_{k_1}}{dt^2} + \omega^2(k_1)V_{k_1} &= \left(\frac{\varepsilon}{2}\right) k_2^2 V_{\Phi_0}^2 e^{-i\omega_0 t} V^*_{k_2} \\ \frac{d^2 V^*_{k_2}}{dt^2} + \omega^2(k_2)V^*_{k_2} &= -\left(\frac{\varepsilon}{2}\right) k_1^2 V_{\Phi_0}^2 e^{i\omega_0 t} V_{k_1}. \end{aligned} \tag{1.2.10}$$

The notation introduced in (1.2.10) was $k_2 \equiv k_0 - k_1$. Hence, the terms in the right-hand side are resonant if $\omega_0 - \omega(k_2) = \omega(k_1)$. Taking into account the relation between wave vectors, we can write that

$$\left.\begin{aligned} \omega_0 &= \omega(k_1) + \omega(k_2) \\ k_0 &= k_1 + k_2. \end{aligned}\right\} \tag{1.2.11}$$

These are the decay conditions, or, in different terms, the conditions of phase and frequency matching.

Now we can rewrite the solution, in accordance with our previous argument, as

$$V_{k_i} = a_i(t) \exp[-i\omega_i(k_i)t]. \tag{1.2.12}$$

[1] Note that degeneration is sometimes possible, with both conditions satisfied. Such cases have been treated in the papers by Karpliuk and Oraevsky (1967), Oraevsky and Tsytovich (1967) and Oraevsky and Pavlenko (1969).

It is now easy to write equations for slowly varying amplitudes $a_i(t)$:

$$\left.\begin{aligned} -2\mathrm{i}\omega_1 \frac{\mathrm{d}a_1}{\mathrm{d}t} &= -\left(\frac{\varepsilon}{2}\right) k_2^2 V_{\Phi_0}^2 a_2^* \,\mathrm{e}^{-\mathrm{i}\Delta\omega t} \\ 2\mathrm{i}\omega_2 \frac{\mathrm{d}a_2^*}{\mathrm{d}t} &= -\left(\frac{\varepsilon}{2}\right) k_1^2 V_{\Phi_0}^2 a_1 \,\mathrm{e}^{\mathrm{i}\Delta\omega t} \\ \Delta\omega = \omega_0 - \omega_1 - \omega_2. & \end{aligned}\right\} \tag{1.2.13}$$

It is now readily seen that the solution can be written as

$$\left.\begin{aligned} a_1 &\sim \exp\left[-\mathrm{i}\frac{\Delta\omega}{2}t + \nu t\right] \\ a_2^* &\sim \exp\left[\mathrm{i}\frac{\Delta\omega}{2}t + \nu t\right] \\ \nu &= \sqrt{\gamma_D^2 - \left(\frac{\Delta\omega}{2}\right)^2} \quad \gamma_D^2 \equiv \frac{\varepsilon^2 k_1^2 k_2^2 V_\Phi^2}{16\omega_1\omega_2}. \end{aligned}\right\} \tag{1.2.14}$$

This solution describes the parametric decay instability. As follows from (1.2.14), if (1.2.11) is exactly satisfied (zero frequency detuning, that is, $\Delta\omega = 0$), the amplitudes of the a_1 and a_2 waves grow exponentially with increment $\nu = \gamma_D$. The product $\omega_1\omega_2$ must then be greater than zero, which leads, together with the decay conditions (1.2.11), to the inequality

$$\omega_0 > \omega_1, \omega_2.$$

In other words, decay instability typically excites 'red' satellites (lower frequencies). 'Blue' satellites are typically excited if degeneracy is present (Karpliuk and Oraevsky 1967, Oraevsky and Tsytovich 1967, Oraevsky and Pavlenko 1969), when the following conditions are simultaneously satisfied:

$$\left.\begin{aligned} \omega_0 - \omega_1 &= \omega_2 \\ k_0 - k_1 &= k_2 \end{aligned}\right\} \quad \text{and} \quad \left.\begin{aligned} \omega_0 + \omega_1 &= \omega_3 \\ k_0 + k_1 &= k_3. \end{aligned}\right\} \tag{1.2.15}$$

PDI thresholds. It is not difficult to take into account the wave dissipation and to find its effect on PDI, by a method similar to that used to find the width of PDI bands. This is easily done by introducing small imaginary increments to natural frequencies. Equations of type (1.2.15) then gain terms $\omega_i + \mathrm{i}\gamma_i$, where γ_i are damping decrements for the corresponding waves. Assuming for simplicity $\Delta\omega = 0$ and performing elementary algebra, we arrive at the following expression for the PDI increment covering dissipation:

$$\nu_D = -\frac{(\gamma_1 + \gamma_2)}{2} + \sqrt{\gamma_D^2 + \frac{(\gamma_1 - \gamma_2)^2}{4}} \tag{1.2.16}$$

which gives an expression for the instability threshold,

$$\gamma_{DT}^2 = \gamma_1\gamma_2 \tag{1.2.17}$$

or, for the problem formulated above with phase velocity modulation,

$$\varepsilon_T^2 = \frac{16\omega_1\omega_2\gamma_1\gamma_2}{k_1^2k_2^2V_\Phi^4}. \tag{1.2.18}$$

In other words, PDI arises when the modulation amplitude ε exceeds the value given by (1.2.18). Note also that (1.2.18) implies that the threshold vanishes if even one of the decrements of the doublet tends to zero.

The discussion above refers to cases in which inhomogeneity produces negligible effects: the analysis was given in the homogeneous medium approximation. In inhomogeneous media, oscillations drift out of the resonant interaction band. This also produces PDI thresholds.

As we know, the invariant characteristic in wave propagation through weakly inhomogeneous media is frequency; the 'quasi-classical' value of the wave vector k_i is found from the equation

$$\omega_i(k_i, x) = \text{const}. \tag{1.2.19}$$

This means that the phase matching is being destroyed (decay conditions for k_i) while frequency matching is sustained. Here we encounter not the temporal but the spatial problem of parametrically coupled oscillators. The set of truncated equations for this case can be obtained by a simple substitution in (1.2.13): $\frac{\partial}{\partial t} \longrightarrow V_{gi}\frac{\partial}{\partial x}$, where $V_{gi} = \frac{d\omega_i}{dk}$ are the corresponding group velocities. It is also necessary to replace $-i\Delta\omega t$ in the exponentials by $i\Delta k x$, where $\Delta k = k_0 - k_1 - k_2$. Now it is not difficult to obtain the spatial increment of PDI, κ_D, instead of the temporal one:

$$\kappa^2 = \frac{\gamma_D^2}{(d\omega_1/dk)\,(d\omega_2/dk)}. \tag{1.2.20}$$

Obviously, equation (1.2.16) (which takes into account the temporal detuning $\Delta\omega$ of the resonance) is now replaced by an expression for the spatial PDI increment κ:

$$\kappa = \sqrt{\kappa_D^2 - \left(\frac{\Delta k}{2}\right)^2}. \tag{1.2.21}$$

Expression (1.2.21) allows us to solve the problem of wave amplification in the interaction band determined by the formula $\Gamma \approx k\Delta x_0$. We see that we need to calculate the length of the interaction zone with amplification Δx_0. It can be found from the condition of saturation of parametric amplification (pumping). As follows from (1.2.21), $\kappa = 0$ when $\kappa_D = \frac{\Delta k}{2}$ and $\Delta k = \frac{d}{dx}(k_0 - k_1 - k_2)\Delta x_0$ is the detuning of the spatial resonance due to inhomogeneity. Simple algebra gives

$$\Delta x_0 \approx 2\gamma_D \Big/ \frac{d}{dx}(k_0 - k_1 - k_2)\sqrt{\frac{d\omega_1}{dk}\frac{d\omega_2}{dk}} \tag{1.2.22}$$

or for Γ,

$$\Gamma \approx 2\gamma_D^2 \Big/ \frac{d\omega_1}{dk} \frac{d\omega_2}{dk} \frac{d}{dx}(k_0 - k_1 - k_2). \tag{1.2.23}$$

1.3 PDI of magnetohydrodynamic waves (quantitative theory)

In this section we outline the quantitative theory of PDI, using as an example a compressible magnetohydrodynamic medium (Oraevsky 1984).

We will recall that the equations of magnetic hydrodynamics have a wide range of applicability. They describe not only electrically conducting gas-dynamic media, in which collision lengths and times are much smaller than the characteristic spatial and temporal parameters of the problem. Chew *et al* (1956), Rudakov and Sagdeev (1958) and Oraevsky *et al* (1968) have shown that these equations can also be applied to collisionless plasma. Magnetic hydrodynamics equations for isotropic pressure are usually written in the form

$$\left.\begin{aligned} &\frac{d\rho}{dt} + \operatorname{div}(\rho \boldsymbol{V}) = 0 \\ &\rho\frac{d\boldsymbol{V}}{dt} = -\nabla P + \frac{1}{4\pi}[\operatorname{rot} \boldsymbol{H}, \boldsymbol{H}] \\ &\frac{d}{dt}\left(\frac{P}{\rho^\gamma}\right) = 0 \\ &\frac{\partial \boldsymbol{H}}{\partial t} = \operatorname{rot}[\boldsymbol{V}\boldsymbol{H}] \end{aligned}\right\} \tag{1.3.1}$$

where $d/dt \equiv (\partial/\partial t) + \boldsymbol{V}\nabla$, ρ is the mass density, p is pressure, $\boldsymbol{V}$ is the mass velocity, $\boldsymbol{H}$ is the magnetic field and γ is the adiabatic exponent. For small-amplitude waves, we can rewrite these quantities as

$$\rho = \rho_0 + \delta\rho \qquad \boldsymbol{V} = \delta\boldsymbol{V} \qquad P = P_0 + \delta P \qquad \boldsymbol{H} = \boldsymbol{H}_0 + \delta\boldsymbol{h}$$

where ρ_0, P_0 and $\boldsymbol{H}_0$ are constants that characterize the state of a uniform 'background' magnetohydrodynamic medium, on which propagate and interact magnetohydrodynamic waves that are described by the quantities $\delta\rho$, $\delta\boldsymbol{V}$, δP, $\delta\boldsymbol{h}$.

We assume that the deviations $\delta\rho$, $\delta\boldsymbol{V}$, δP, $\delta\boldsymbol{h}$ from the 'background' are fairly small and therefore retain only the terms up to and including those quadratic in deviations. Terms of third and higher orders are omitted. In this case the system of equations for the deviations from the 'background' takes the form

$$\partial(\delta\rho)/\partial t + \rho_0 \operatorname{div}(\delta\boldsymbol{V}) = -\operatorname{div}(\delta\rho\,\delta\boldsymbol{V}) \tag{1.3.2}$$

$$\frac{\partial(\delta\boldsymbol{V})}{\partial t} + s^2\frac{\nabla(\delta\rho)}{\rho_0} - \frac{1}{4\pi\rho_0}[\operatorname{rot}(\delta\boldsymbol{h})\boldsymbol{H}_0] = -(\delta\boldsymbol{V}\nabla)\delta\boldsymbol{V} - s^2(\gamma-2)\frac{\delta\rho\nabla\,\delta\rho}{\rho_0^2}$$

$$
+\frac{1}{4\pi\rho_0}[\operatorname{rot}(\delta\boldsymbol{h})\,\delta\boldsymbol{h}]-\frac{1}{4\pi\rho_0}[\operatorname{rot}(\delta\boldsymbol{h})\boldsymbol{H}_0] \tag{1.3.3}
$$

$$
\frac{\partial(\delta\boldsymbol{h})}{\partial t}-\operatorname{rot}[\delta\boldsymbol{V}\boldsymbol{H}_0]=\operatorname{rot}[\delta\boldsymbol{V}\,\delta\boldsymbol{h}] \tag{1.3.4}
$$

where $s^2 \equiv \gamma\, p_0/\rho_0$ is the velocity of sound squared.

In the case of homogeneous plasma, it is convenient to pass from the quantities $\delta\rho$, $\delta\boldsymbol{V}$ and $\delta\boldsymbol{H}$ to their Fourier transforms in spatial variables ρ_k, $\boldsymbol{V}_k$ and $\boldsymbol{h}_k$. This gives rise to (for (1.3.2)–(1.3.4))

$$
\mathrm{i}\frac{\partial\rho_k}{\partial t}-\rho_0\boldsymbol{k}\boldsymbol{V}_k=\sum_{k_1+k_2=k}\rho_{k_1}\boldsymbol{k}\boldsymbol{V}_{k_2} \tag{1.3.5}
$$

$$
\begin{aligned}
\mathrm{i}\frac{\partial\boldsymbol{V}_k}{\partial t}-\boldsymbol{k}s^2\frac{\rho_k}{\rho_0}+\frac{1}{4\pi\rho_0}[[\boldsymbol{k}\boldsymbol{h}_k]\boldsymbol{H}_0]=\sum_{k_1+k_2=k}\Big\{(\boldsymbol{V}_{k_1}\boldsymbol{k}_2)\boldsymbol{V}_{k_2}\\
+\frac{s^2(\gamma-2)}{\rho_0^2}\boldsymbol{k}_2\rho_{k_1}\rho_{k_2}-\frac{1}{4\pi\rho_0}[[\boldsymbol{k}_1\boldsymbol{h}_{k_1}]\boldsymbol{h}_{k_2}]+\frac{\rho_{k_1}}{4\pi\rho_0^2}[[\boldsymbol{k}_2\boldsymbol{h}_{k_2}]\boldsymbol{H}_0]\Big\}
\end{aligned} \tag{1.3.6}
$$

$$
\mathrm{i}\frac{\partial\boldsymbol{h}_k}{\partial t}+[\boldsymbol{k}[\boldsymbol{V}_k\boldsymbol{H}_0]]=-\sum_{k_1+k_2=k}[\boldsymbol{k}[\boldsymbol{V}_{k_1}\boldsymbol{h}_{k_2}]]. \tag{1.3.7}
$$

In matrix form these equations can be rewritten as

$$
\mathrm{i}\frac{\partial\Psi_\alpha}{\partial t}-\hat{H}^{(0)}_{\alpha\beta}\Psi_\beta=U^{(1)}_\alpha\left\{\sum_{k=k_1+k_2}(\Psi_{k_1},\Psi_{k_2})\right\}. \tag{1.3.8}
$$

The components of the column vector Ψ_α in (1.3.8) are the mass density and the projections of the mass velocity and magnetic field, that is

$$
\Psi_1\equiv\rho_k \qquad \Psi_{2,3,4}\equiv V^{\boldsymbol{k}}_{x,y,z} \qquad \Psi_{5,6,7}\equiv h^{\boldsymbol{k}}_{x,y,z}. \tag{1.3.9}
$$

In the coordinate frame of reference in which the z-axis points along the magnetic field $\boldsymbol{H}$, the matrix $H^0_{\alpha\beta}$ has the form

$$
\left\|\begin{array}{ccccccc}
0 & k_x\rho_0 & k_y\rho_0 & k_z\rho_0 & 0 & 0 & 0\\
k_x\frac{s^2}{\rho_0} & 0 & 0 & 0 & -\frac{k_zV_A}{\sqrt{4\pi\rho_0}} & 0 & \frac{k_xV_A}{\sqrt{4\pi\rho_0}}\\
k_y\frac{s^2}{\rho_0} & 0 & 0 & 0 & 0 & -\frac{k_zV_A}{\sqrt{4\pi\rho_0}} & \frac{k_yV_A}{\sqrt{4\pi\rho_0}}\\
k_z\frac{s^2}{\rho_0} & 0 & 0 & 0 & 0 & 0 & 0\\
0 & -k_zH_0 & 0 & 0 & 0 & 0 & 0\\
0 & 0 & -k_zH_0 & 0 & 0 & 0 & 0\\
0 & k_xH_0 & k_yH_0 & 0 & 0 & 0 & 0
\end{array}\right\| \tag{1.3.10}
$$

where $V_A = H_0/\sqrt{4\pi\rho_0}$ is the Alfven velocity.

The eigenvectors of the operator $\hat{H}^{(0)}_{\alpha\beta}$ naturally describe linear magnetohydrodynamic waves. Indeed, in the linear approximation $U^{(1)}_\alpha$ must be set to zero. Equation (1.3.8) then takes the form[2]

$$\frac{\mathrm{i}\partial\Psi_\alpha}{\partial t} - \hat{H}^{(0)}_{\alpha\beta}\Psi_\beta = 0. \qquad (1.3.11)$$

The solution to (1.3.11) can be sought as a Fourier expansion in time. It has the form

$$\omega\Psi_\alpha - H^{(0)}_{\alpha\beta}\Psi_\beta = 0. \qquad (1.3.12)$$

The condition of solvability of this equation gives the eigenfrequencies of the problem. In other words, we obtain the familiar dispersion equation for MHD waves,

$$(\omega^2 - k_z^2 V_A^2)\left\{\omega^4 - \omega^2 k^2 (V_A^2 + s^2) + k^2 k_z^2 V_A^2 s^2\right\} = 0. \qquad (1.3.13)$$

For the sake of simplicity, we consider here low-pressure plasma in which $\beta \equiv p_0/(H^2/8\pi) \ll 1$. In this case, the solutions to (1.3.13) take a simpler form,

$$\omega_a = \pm k_z V_A \qquad \omega_m = \pm k V_A \qquad \omega_s = \pm k_z s \qquad (1.3.14)$$

(equations (1.3.12)–(1.3.14) allow one to find the eigenvectors of the problem; it is not difficult to see that these eigenvectors describe the Alfven (ω_a), the fast (ω_m) and the slow (ω_s) magnetoacoustic waves). By substituting each eigenvalue of (1.3.14) into (1.3.12) and solving the resulting set of equations, we obtain

$$\Psi^a_\alpha = c^a e^a_\alpha \exp(-\mathrm{i}\omega_a t) \qquad \Psi^{m,s}_\beta = c^{m,s} e^{m,s}_\beta \exp(-\mathrm{i}\omega_{m,s} t) \qquad (1.3.15)$$

$$e^a(k,\omega) = \sqrt{\frac{|\omega_a|}{\rho_0}}\left\|\begin{matrix} 0 \\ k_y/k_\perp \\ k_x/k_\perp \\ 0 \\ k_y k_z H_0/k_\perp\omega_a \\ k_x k_z H_0/k_\perp\omega_a \\ 0 \end{matrix}\right\| \qquad e^m = \sqrt{\frac{|\omega_m|}{\rho_0}}\left\|\begin{matrix} k_\perp\rho_0/kV_A \\ k_x k V_A/k_\perp\omega_m \\ k_y k V_A/k_\perp\omega_m \\ 0 \\ -k_x k_z H_0/k_\perp k V_A \\ -k_y k_z H_0/k_\perp k V_A \\ k_\perp H_0/kV_A \end{matrix}\right\|$$

$$e^s = \sqrt{\frac{|\omega_s|}{\rho_0}}\left\|\begin{matrix} \rho_0/s \\ 0 \\ 0 \\ k_s s/\omega_s \\ -k_x k_z s H_0/k^2 V_A^2 \\ k_y k_z s H_0/k^2 V_A^2 \\ -k_\perp^2 s H_0/k^2 V_A^2 \end{matrix}\right\|. \qquad (1.3.16)$$

The vectors e^a, e^m and e^s are not orthogonal under the standard definition of the scalar product, under which $\langle \boldsymbol{x}, \boldsymbol{y}\rangle = \sum x_\alpha y_\alpha$. However, they can be

[2] The form of equation (1.3.11) is similar to that of the Schrödinger–Dirac-type quantum-mechanical equations. This fact has profound consequences.

orthogonalized by standard techniques, by defining the scalar product as[3]

$$\langle \bar{\boldsymbol{x}}, \bar{\boldsymbol{y}} \rangle \equiv \sum_{\alpha,\beta=1}^{7} x_\alpha^* \gamma_{\alpha\beta}^{(0)} y_\beta. \tag{1.3.17}$$

The matrix $\gamma_{\alpha\beta}^{(0)}$ is a matrix which transforms the eigenvectors $\boldsymbol{e}^\sigma$ of the operator $\hat{H}^{(0)}$ to the eigenvectors $\bar{\boldsymbol{e}}^\sigma$ of the conjugate operator $\tilde{\hat{H}}^{(0)}$, that is

$$\bar{e}_\beta^\sigma = \gamma_{\beta\alpha} e_\alpha^\sigma \tag{1.3.18}$$

where

$$\gamma_{\beta\alpha} = \begin{Vmatrix} \frac{s}{2\rho_0} & 0 & 0 & 0 & 0 & 0 & 0 \\ 0 & \frac{\rho_0}{2} & 0 & 0 & 0 & 0 & 0 \\ 0 & 0 & \frac{\rho_0}{2} & 0 & 0 & 0 & 0 \\ 0 & 0 & 0 & \frac{\rho_0}{2} & 0 & 0 & 0 \\ 0 & 0 & 0 & 0 & \frac{1}{8\pi} & 0 & 0 \\ 0 & 0 & 0 & 0 & 0 & \frac{1}{8\pi} & 0 \\ 0 & 0 & 0 & 0 & 0 & 0 & \frac{1}{8\pi} \end{Vmatrix}. \tag{1.3.19}$$

It can be easily shown that

$$\langle \boldsymbol{e}^\sigma, \boldsymbol{e}^\delta \rangle \equiv e_\alpha^{\sigma *} \gamma_{\alpha\beta} e_\beta^\delta = \delta^{\sigma\delta} |\omega^\sigma| \tag{1.3.20}$$

where $\delta^{\sigma\delta}$ is the Kronecker symbol.

Obviously, the matrix $\gamma_{\alpha\beta}^{(0)}$ is defined up to a constant factor.

This factor is chosen in (1.3.19) in such a way that the eigenvector is normalized to the energy of the relevant MHD wave $|\Psi^\sigma|^2 = |c_0|^2 |\omega^\sigma| = \varepsilon^\sigma$. By introducing the wave amplitude c^σ in this way, we can interprete $|c^\sigma|^2$ as the number of quasi-particles (MHD waves) in the state σ because

$$|c_\sigma|^2 \equiv N^\sigma = \varepsilon^\sigma / |\omega^\sigma|. \tag{1.3.21}$$

It is now easy to obtain equations that describe the evolution of amplitudes c^σ of the MHD waves owing to their nonlinear interaction. To achieve this, we return to equation (1.3.8). It follows from (1.3.5)–(1.3.7) that $U_\alpha^{(1)}$ in (1.3.8) has the form

$$\begin{aligned} U_1^{(1)} &= \sum_{\boldsymbol{k}_1+\boldsymbol{k}_2=\boldsymbol{k}} \rho_{\boldsymbol{k}_1} \boldsymbol{k} \boldsymbol{V}_{\boldsymbol{k}_2} \\ U_{2.3.4}^{(1)} &= \Bigg\{ \sum_{\boldsymbol{k}_1+\boldsymbol{k}_2=\boldsymbol{k}} (\boldsymbol{V}_{\boldsymbol{k}_1} \boldsymbol{k}_2) \boldsymbol{V}_{\boldsymbol{k}_2} + \frac{s^2(\gamma-2)}{\rho_0^2} \rho_{\boldsymbol{k}_1} \rho_{\boldsymbol{k}_2} \boldsymbol{k}_1 \\ &\quad - \frac{1}{4\pi\rho_0} [[\boldsymbol{k}_1 \boldsymbol{h}_{\boldsymbol{k}_1}] \boldsymbol{h}_{\boldsymbol{k}_2}] + \frac{\rho_{\boldsymbol{k}_1}}{4\pi\rho_0^2} [[\boldsymbol{k}_2 \boldsymbol{h}_{\boldsymbol{k}_2}] \boldsymbol{H}_0] \Bigg\}_{x,y,z} \\ U_{5.6.7}^{(1)} &= [\boldsymbol{k}(\boldsymbol{V}_{\boldsymbol{k}_1}, \boldsymbol{h}_{\boldsymbol{k}_2})]_{x,y,z}. \end{aligned} \tag{1.3.22}$$

[3] Note that a similar definition of the scalar product is given in relativistic quantum mechanics for the Dirac equation.

Using (1.3.8) and (1.3.23), it is not difficult to write equations for the amplitudes c^α. To achieve this, we expand, as is usual, the total vector $\Psi(\boldsymbol{k}, t)$ that describes the state of the MHD medium in powers of the eigenvectors of the linear problem:

$$\Psi(\boldsymbol{k}, t) = \sum_\sigma c^\sigma \exp(-\mathrm{i}\omega^\sigma t)\, \boldsymbol{e}^\sigma. \tag{1.3.23}$$

Substituting (1.3.23) into (1.3.8) (taking into account (1.3.9)) and multiplying by the eigenvector $\boldsymbol{e}^\alpha$, we obtain the following set of equations for amplitudes of the waves c^α that describe the interaction of these waves,

$$\mathrm{i}\frac{\partial c^\alpha(\boldsymbol{k}, t)}{\partial t} = \sum_{\substack{\boldsymbol{k}_1+\boldsymbol{k}_2=\boldsymbol{k}\\ \alpha_1,\alpha_2}} V_{\boldsymbol{k},\alpha,\boldsymbol{k}_1,\alpha_1,\boldsymbol{k}_2,\alpha_2} c^{\alpha_1}(\boldsymbol{k}_1, t) c^{\alpha_2}(\boldsymbol{k}_2, t) \times \exp[-\mathrm{i}(\omega_\alpha - \omega_{\alpha_1} - \omega_{\alpha_2})t] \tag{1.3.24}$$

where

$$V_{\boldsymbol{k},\alpha,\boldsymbol{k}_1,\alpha_1,\boldsymbol{k}_2,\alpha_2} = \frac{\langle \boldsymbol{e}^\alpha(\boldsymbol{k}_1), U^{(1)}(\boldsymbol{e}^{\alpha_1}(\boldsymbol{k}_1), \boldsymbol{e}^{\alpha_2}(\boldsymbol{k}_2))\rangle}{|\omega_\alpha(\boldsymbol{k})|}. \tag{1.3.25}$$

Now it is not difficult to single out, as in the preceding section, only the resonant terms. We readily see that the MHD waves creating the resonance contribution are tied to the α wave by the conditions

$$\boldsymbol{k}^\alpha = \boldsymbol{k}_1^{\alpha_1} + \boldsymbol{k}_2^{\alpha_2} \qquad \omega^\alpha = \omega_1^{\alpha_1} + \omega_2^{\alpha_2}. \tag{1.3.26}$$

Henceforth we omit the indices α that characterize the type and polarization of waves. In plasma physics, conditions (1.3.26) are referred to as decay conditions (Oraevsky and Sagdeev 1962), and in nonlinear radiophysics and optics as conditions of phase and frequency matching (Kroll 1962). Equations (1.3.24) are often called the equations for coupled modes, and $V_{\boldsymbol{k},\boldsymbol{k}_1,\boldsymbol{k}_2}$ are known as matrix elements or mode coupling coefficients.

It is important to note that $V_{\boldsymbol{k},\boldsymbol{k}_1,\boldsymbol{k}_2}$ have the symmetries first pointed out by Kroll (1962):

$$\left.\begin{aligned} &V_{\boldsymbol{k},\boldsymbol{k}_1,\boldsymbol{k}_2} = V_{\boldsymbol{k},\boldsymbol{k}_2,\boldsymbol{k}_1} \qquad V_{\boldsymbol{k},\boldsymbol{k}_1,\boldsymbol{k}_2} = V_{-\boldsymbol{k},-\boldsymbol{k}_1,-\boldsymbol{k}_2} \\ &V_{\boldsymbol{k},\boldsymbol{k}_1,\boldsymbol{k}_2} = V_{\boldsymbol{k}_2,-\boldsymbol{k}_1,\boldsymbol{k}} \operatorname{sgn}[\omega_1(\boldsymbol{k}_1)\cdot\omega_2(\boldsymbol{k}_2)]. \end{aligned}\right\} \tag{1.3.27}$$

The first of these conditions is fairly obvious: it follows from the symmetry under permutation of the waves $c(\boldsymbol{k}_1)$ and $c(\boldsymbol{k}_2)$. The second condition is implied by the fact that the density, mass velocity and magnetic field that describe MHD waves are real quantities. The third condition in (1.3.27) is valid only for wave triplets coupled by the decay conditions (1.3.26). It can be treated as a consequence of the laws of energy and momentum conservation in three-wave interactions. Indeed, in view of (1.3.24) and (1.3.26), we can obtain

$$(\partial/\partial t)(N_k|\omega_k| + N_{k_1}|\omega_{k_1}| + N_{k_2}|\omega_{k_2}|)$$
$$= 2\sin\theta\sqrt{N_k N_{k_1} N_{k_2}}\,[V_{k,k_1,k_2}\,\mathrm{sgn}(\omega_k)\omega_k - V_{k_1,k,-k_2}\,\mathrm{sgn}(\omega_{k_1})\omega_{k_1}$$
$$- V_{-k_2,-k,k_1}\,\mathrm{sgn}(\omega_{k_2})\omega_{k_2}] \qquad (1.3.28)$$
$$\theta = \varphi_k - \varphi_{k_1} - \varphi_{k_2} \qquad c_k = \sqrt{N_k}\,\exp(\mathrm{i}\varphi_k) \qquad \mathrm{Im}\,\varphi_k = 0.$$

The derivative in the left-hand side of (1.3.28) is applied to the sum of energies of three interacting waves. This means that the left-hand side vanishes. Indeed, the right-hand side of (1.3.28) becomes zero as well, regardless, of course, of the value of the relative phase θ. Substituting the quantities from (1.3.6) into (1.3.28), it is not difficult to derive the third symmetry property for the mode coupling coefficients in (1.3.27).

We have carried out above a quantitative analysis for MHD waves. It is quite clear that all qualitative conclusions drawn remain valid for any nonlinear media. The reason for this is that if wave damping is neglected, we deal with Hermitian operators for linear waves. The truncated set of equations (1.3.24) is therefore unchanged for any type of nonlinear wave-sustaining media. The medium type is reflected in the coupling coefficients V_{k,k_1,k_2} between the waves. Their symmetry properties are equally independent of a specific medium.

Set (1.3.24) allows one to describe the decay instability in any nonlinear medium in a sufficiently general form. As a first step, we can ignore the effect of the fluctuation waves c_1 and c_2 in the first approximation for waves of finite but moderately high amplitude; c_1 and c_2 are related via (1.3.26) with the initial wave c_0 known as the pump wave. Assuming that the pump wave is constant, we arrive at a set of equations describing the time-dependent amplitudes of the waves c_1 and c_2:

$$\partial c_1(\omega_1, k_1, t)/\partial t = V_{k_1,k_0,-k_2} c_0 c_2(-\omega_2, -k_2, t)$$
$$\partial c_2(-\omega_2, -k_2, t)/\partial t = V_{-k_2,-k_0,k_1} c_0(-\omega, -k) c_1(\omega_1, k_1, t). \qquad (1.3.29)$$

Equation (1.3.29) is easily solved. Taking into account that the amplitudes $c_i(-\omega_i, -k_i) = c_i^*(\omega_i, k_i)$ are real and also the symmetric properties of wave coupling coefficients indicated in (1.3.27), we find

$$c_1(\omega_1, k_1, t) = c_1^-(\omega_1, k_1, 0)\exp(-\gamma_D t) + c_1^+(\omega_1, k_1, 0)\exp(\gamma_D t) \qquad (1.3.30)$$

$$c_2(-\omega_2, -k_2, t) = c_2^-(-\omega_2, -k_2, 0)\exp(-\gamma_D t) + c_2^+(-\omega_2, -k_2, 0)\exp(\gamma_D t) \qquad (1.3.31)$$

where $\gamma_D^2 = |V_{k_1,k_0,-k_2}|^2|c_0|^2 > 0$.

A remark is in order here. Even though the resonant interaction occurs also for waves related to the initial wave by the conditions

$$\omega_0 = \omega_2 - \omega_1 \qquad k_0 = k_2 - k_1$$

no instability occurs for this triplet of waves. Indeed, we have in this case

$$\gamma_D^2 = -|V_{\boldsymbol{k}_1,\boldsymbol{k}_0,\boldsymbol{k}_2}|^2|c_0|^2 < 0.$$

What we encounter here is not an instability but a frequency shift for the waves c_1 and c_2.

An interesting specific example of decay instability in magnetic hydrodynamics is readily available. It is well known that the Alfven sine waves in magnetic hydrodynamics are an exact solution of nonlinear equations (owing to the polarization and incompressibility of such types of MHD motion, no sharpening and anharmonism arise for such waves). It is not surprising, therefore, that it has been assumed for a long time that Alfven waves can propagate infinitely long in zero dissipation media without changing form. It is not difficult to show, however, that Alfven waves are unstable under perturbations that must contain, in addition to the Alfven waves, magnetoacoustic waves as well. Indeed, decay conditions (1.3.26) can be satisfied for the wave triplets

$$A \to A_1 + S_2 \qquad A \to M_1 + S_2 \qquad A \to A_1 + M_2$$

where A, M and S stand for the Alfven, slow and fast magnetoacoustic waves, respectively. In low-pressure plasmas ($\beta \ll 1$) the increment is maximum for the instability due to the excitation of the doublet formed by the new Alfven wave (A_2) and the slow magnetoacoustic wave. Calculating the corresponding coupling coefficient $V_{A_1,A_0,-S_2}$ as shown in the preceding analysis, we obtain an expression for the increment of the instability considered (Galeev and Oraevsky 1962):

$$\gamma_D = \left\{ \frac{k_{2y}^2 \delta V_A^2}{16} \left[1 + \left(\frac{k_{2z} k_{2y} s^2}{\omega^2 - k_{2z}^2 s^2} \right)^2 \right] \right\}$$
$$\times \left\{ \frac{\omega_2^2 \cos\delta}{\omega_2^2 - k_{2z}^2 s^2} - 4 \sin\gamma \sin(\gamma + \delta) \right\} \frac{\omega_1}{\omega_2} \qquad (1.3.32)$$

where δ and γ are the angles between the planes $\boldsymbol{k}_0$, $\boldsymbol{H}_0$ and $\boldsymbol{k}_1$, $\boldsymbol{H}_1$, and also $\boldsymbol{k}_1$, $\boldsymbol{H}_0$ and $\boldsymbol{k}_2$, $\boldsymbol{H}_0$, respectively; the y-axis is chosen perpendicularly to $\boldsymbol{H}_0$ in the plane $\boldsymbol{k}_2$, $\boldsymbol{H}_0$. Clearly, this increment is of the order of $(\delta V_A/\sqrt{8 V_A S})\omega_0$. If instead of the Alfven wave, the perturbation contains the fast magnetoacoustic wave, the increment is somewhat lower but of the same order of magnitude as (1.3.32). The increments of instabilities of other types of instability are much lower than (1.3.32). We emphasize that the instability itself only arises if compressibility is taken into account (magnetoacoustic waves). This gives an important conclusion that if plasma compressibility is neglected, turbulence spectra of Alfven waves cannot be correctly calculated, especially for low-pressure plasmas ($\beta \ll 1$).

1.4 Modulation instability

The modulation instability is a frequently encountered form of direct-type nonlinear instability. This instability was first considered by Vedenov and Rudakov (1964) and Lighthill (1965). The instability criterion was first derived by Lighthill (1965). To find it, consider the general equation of energy transfer. Note first of all that the main nonlinear contribution for low-amplitude waves is proportional to frequency squared. This is the well known effect of nonlinear self-action and thus will not be discussed here in detail. The expression for frequency can be written in the form

$$\omega(k, a) = \omega_0(k) + \alpha|a|^2 \tag{1.4.1}$$

where $\omega_0(k)$ is the wave frequency in the linear approximation and $\alpha|a|^2$ is the nonlinear frequency shift. We see that α is the proportionality coefficient that relates the nonlinear frequency shift to the squared amplitude of the pump wave. Following Kadomtsev and Karpman (1971), we can rewrite the energy transfer equation in the general form

$$\frac{\partial a^2}{\partial t} + \frac{\partial}{\partial x}(V_g a^2) = 0 \tag{1.4.2}$$

(in this equation and hereafter we assume the wave amplitude to be real). We need to add to equation (1.4.2) an equation relating changes in time of the quasi-momentum $k \equiv \frac{\partial \varphi}{\partial x}$ of the wave with changes in time of the quasi-energy $\omega \equiv \frac{\partial \varphi}{\partial t}$, where φ is the phase of the wave. It is fairly obvious that this equation is

$$\frac{\partial k}{\partial t} = -\frac{\partial \omega}{\partial x}. \tag{1.4.3}$$

Substituting into (1.4.3) the expression for frequency (1.4.1), we obtain

$$\frac{\partial k}{\partial t} = -V_g \frac{\partial k}{\partial x} - \alpha \frac{\partial a^2}{\partial x}. \tag{1.4.4}$$

Let low-amplitude modulation arise on a monochromatic wave with amplitude a_0 and wavenumber k_0,

$$\left.\begin{aligned} a &= a_0 + \delta a \exp(-\mathrm{i}\Omega t + \mathrm{i}\kappa x) \\ k &= k_0 + k' \exp(-\mathrm{i}\Omega t + \mathrm{i}\kappa x) \end{aligned}\right\}$$

assuming here $\Omega \ll \omega_0$, $\kappa \ll k_0$.

By linearizing equations (1.4.2) and (1.4.4), we easily derive the dispersion equation relating Ω to κ:

$$\Omega = \kappa V_g \pm \sqrt{\alpha \left(\frac{\partial V_g}{\partial k}\right) \kappa a^2}. \tag{1.4.5}$$

Equation (1.4.5) implies in an obvious manner the instability condition:

$$\alpha \frac{\partial V_g}{\partial k} < 0. \tag{1.4.6}$$

This is the Lighthill criterion. The physical interpretation of the criterion is readily formulated. To be specific, assume $\alpha > 0$. Then the phase velocity in the regions of maximum amplitude is higher than in the region of minimal amplitude. In its turn, this implies that the number of nodes increases as we go nearer the region of maximum amplitude and decreases away from it. Hence, if the derivative of the group velocity with respect to k is negative, oscillations approaching the maximum amplitude region fall behind while those receding from it are gaining, thereby amplifying the growth of amplitude at maxima and making the minimum deeper. This is the phenomenon known as the modulational instability. Similar arguments can be developed for the case of $\alpha < 0$.

1.5 Explosive instability of negative-energy waves

Waves are said to possess negative energy because, with the energy of the medium remaining constant, an increase in the wave amplitude reduces the total energy of the medium + wave system. These waves are known both in hydrodynamics and in plasma physics (Landau and Lifshits 1951, Kadomtsev *et al* 1964). To answer the question 'When do such wave arise?', we need to look at the expression for the total energy of the wave. As is well known for electromagnetic waves, the total wave energy is

$$W = \frac{1}{8\pi}\left[\left(\frac{\mathrm{d}}{\mathrm{d}\omega}\right)(\omega\varepsilon)\langle E^2\rangle + \left(\frac{\mathrm{d}}{\mathrm{d}\omega}\right)(\mu\omega)\langle H^2\rangle\right] \tag{1.5.1}$$

where ε and μ are the dielectric and magnetic permittivities, respectively, of the medium, and angle brackets indicate averaging over oscillations. For the sake of simplicity, we consider electrostatic waves. In this case the sign of the energy of the wave is dictated by the sign of $\frac{\mathrm{d}\varepsilon}{\mathrm{d}\omega}$ (since the dispersion equation for the longitudinal (electrostatic) waves is $\varepsilon(\omega, k) = 0$). On the other hand, the sign of $\frac{\mathrm{d}\varepsilon}{\mathrm{d}\omega}$ in a dynamic equilibrium medium is positive. This is a direct corollary of the Kramers–Krönig relations. Therefore, negative-energy waves can arise only in thermodynamically non-equilibrium systems. It is not difficult to show that in an anisotropic plasma traversed by a monoenergetic beam, negative-energy waves are the electrostatic waves whose velocities are close to the beam velocity. Another example is found in ion-cyclotron waves in a magnetoactive plasma at a sufficiently high temperature anisotropy, when $T_\parallel/T_\perp \to 0$. The dielectric permittivity of waves with frequencies close to the ion-cyclotron frequency is given by the expression (Dikasov *et al* 1965)

$$\varepsilon = 1 - \frac{\omega_p^2 k_\parallel^2}{\omega^2 k^2} - \frac{\Omega_p^2 \Gamma_m k_\parallel^2}{(\omega - m\Omega_H)k^2} \tag{1.5.2}$$

where ω_p and Ω_p are the electron and ion plasma frequencies, Ω_H is the ion-cyclotron frequency, $k_\parallel$ is the component of the wave vector k parallel to the magnetic field,

$$\Gamma_m = I_m(b_\perp)\exp(-b_\perp) \qquad b_\perp = k_\perp^2 T/M\Omega_H$$

and I_m is the modified Bessel function of order m.

It is not difficult to show that waves with frequencies sufficiently close to the cyclotron frequencies are negative-energy waves.

Negative-energy waves possess a number of specific properties. For example, taking dissipation into account results not in damping of these waves but in autoexcitation. Indeed, dissipation must increase the energy of the medium while the total energy of the medium + wave system is conserved. Therefore, only the appropriate enhancement of the negative-energy wave can compensate for the increase in the energy of the medium. Let us emphasize that it is unimportant how the energy of the negative-energy wave is dissipated: via linear processes or through nonlinear channeling into positive-energy modes. The logic remains unchanged. This means that these modes are nonlinearly unstable. In order to describe a system of waves interacting with negative-energy waves, it is necessary to apply a method described in section 1.2. It can be readily shown that the following equations are required for interacting modes with arbitrary sign of energy, coupled by decay conditions in low-dissipation media:

$$\begin{aligned} s_i\frac{\partial c_i}{\partial t} &= -\gamma_i c_i + \mathrm{i}\sum_{jk} V_{ijk}c_j c_k \\ s_i &= \operatorname{sgn}\left(\left.\frac{\partial \varepsilon}{\partial \omega}\right|_{\omega=\omega_i}\right) \end{aligned} \tag{1.5.3}$$

The validity of our argument on the instability of waves with linear damping immediately follows from (1.5.3). This instability results in exponential growth of waves. The situation is different for parametrically interacting positive-energy waves which propagate against the 'background' of the negative-energy pump wave. For such a wave triplet, all three waves grow simultaneously. This phenomenon was first described using the kinetic equation which correctly takes into account interactions of waves with random phases and different signs of energy (Dikasov *et al* 1965). Later Timofeev (1966) gave an analysis of cyclotron emission in experiments on plasma confinment in mirror confinement systems. Timofeev traced this radiation to negative-energy cyclotron waves, pointed to the explosive type of the instability and estimated the 'explosion' time. A consistent theory of explosive instability, showing that in the three-wave approximation the amplitudes of all three waves tend to infinity over a finite time ('explosion' time), was given by Coppi *et al* (1969) and Karpliuk *et al* (1970).

The presentation below follows these papers. If linear dissipation is neglected, the equations for the decay mode of three arbitrary-energy waves take the form

$$\left.\begin{aligned} \frac{\partial c_k}{\partial t} &= -\mathrm{i}s_k V_{k,k_1,k_2} c_{k_1} c_{k_2} \\ \frac{\partial c_{k_1}}{\partial t} &= -\mathrm{i}s_{k_1} V_{k,k_1,-k_2} c_k c_{k_2}^* \\ \frac{\partial c_{k_2}}{\partial t} &= -\mathrm{i}s_{k_2} V_{k,-k_1,k_2} c_k c_{k_1}^* \end{aligned}\right\} . \tag{1.5.4}$$

To be specific, we assume the wave with c_k to be the negative-energy pump wave. Introducing new variables u_i and φ_i by $c_{k_i} = u_i \exp(\mathrm{i}\varphi_i)$, we can easily derive from (1.5.4) the following set of equations:

$$\begin{aligned} \frac{\partial u_0}{\partial t} &= V u_1 u_2 \cos\Phi \\ \frac{\partial u_1}{\partial t} &= V u_0 u_2 \cos\Phi \\ \frac{\partial u_2}{\partial t} &= V u_0 u_1 \cos\Phi \\ \frac{\partial \Phi}{\partial t} &= -V\left(\frac{u_1u_2}{u_0} + \frac{u_0u_1}{u_2} + \frac{u_0u_2}{u_1}\right)\sin\Phi \end{aligned} \tag{1.5.5}$$

where $V = |V_{k,k_1,k_2}|$ and $\Phi = \varphi_0 - \varphi_1 - \varphi_2 + \pi/2$. The integrals of motion of (1.5.5) immediately follow:

$$\left.\begin{aligned} m_1 &= u_0^2 - u_1^2 = \text{const} \\ m_2 &= u_0^2 - u_2^2 = \text{const}_2 \\ m_0 &= u_1^2 - u_2^2 = \text{const}_3 \end{aligned}\right\} . \tag{1.5.6}$$

For the system under discussion, these integrals are Manly–Row-type relations. Using the first three equations of (1.5.5), we can transform the last one to the form

$$\frac{\partial \Phi}{\partial t} = \tan\Phi \ln(u_0u_1u_2). \tag{1.5.7}$$

In view of the integrals of motion (1.5.6), the first equation in (1.5.5) can be recast to

$$2Vt = \int_{u_0^2(0)}^{u_0^2(t)} \frac{\mathrm{d}u_0^2}{\sqrt{u_0^2(u_0^2 - m_1)(u_0^2 - m_2)}}. \tag{1.5.8}$$

Assuming, in order to be specific, $u_2^2 > m_1 - m_2$ and changing the variable to $y(t) = m_1^{1/2}/u$, we can transform (1.5.8) to an elliptic integral

$$V\sqrt{m_1}t = -\int_{y(0)}^{y(t)} \frac{\partial y}{\sqrt{(1-y^2)(1-\kappa^2 y)}} \tag{1.5.9}$$

where $\kappa = (m_2/m_1)^{1/2}$. This formula implies that

$$u_0(t) = \frac{m_1^{1/2}}{m_1^{1/2}/u_0(0) - \sin(V_\kappa \sqrt{m_1} t)}. \qquad (1.5.10)$$

Since $u_0 > m_1^{1/2}$, this means that as time tends to some value t_e, the amplitude $u_0 \to \infty$ as $(t - t_e)^{-1}$; hence, the amplitudes u_1 and u_2 tend to infinity following the same law in accordance with (1.5.6). The instability is therefore of an explosive nature.

An inescapable question arising after the type of instability in the three-wave approximation has been determined is: 'How is the explosive instability stabilized?'. This aspect cannot be discussed in any detail within this book. We will only mention that, as shown by Oraevsky *et al* (1973) and Davydova and Oraevsky (1974), the instability is stabilized in different situations either by disturbing the parametric resonance owing to the enhancement of frequency shift in proportion to the amplitude squared of growing waves (violation of frequency matching) or by squeezing the waves in an inhomogeneous medium (violation of phase matching) from the resonance interaction zone.

References

Artsimovich L A and Sagdeev R Z 1979 *Fizika Plazmy dlia Fizikov* (Moscow: Atomizdat) pp 1–320 (in Russian)

Chew G, Goldberger M and Low F 1956 *Proc. R. Soc.* **236** 112

Coppi B, Rosenbluth M N and Sudan R N 1969 *Ann. Phys.* **55** 207

Davydova T A and Oraevsky V N 1974 *Zh. Exp. Teor. Fiz.* **66** 1613

Dikasov V M, Rudakov L I and Ryutov D D 1965 *Zh. Exp. Teor. Fiz.* **48** 913–20

Galeev A A and Karpman V I 1963 *Sov. Phys.–JETP* **17** 403

Galeev A A and Oraevsky V N 1962 *Sov. Phys.–Dokl.* **7** 988

Kadomtsev B B and Karpman V I 1971 *Usp. Fiz. Nauk* **103** 193

Kadomtsev B B, Mikhailovsky A B and Timofeev A V 1964 *Zh. Exp. Teor. Fiz.* **47** 2266–8

Karpliuk K S and Oraevsky V N 1967 *Pisma JETP* **5** 451

Karpliuk K S, Oraevsky V N and Pavlenko V P 1970 *Ukr. Fiz. Zh.* **15** 340

Kroll N 1962 *Phys. Rev.* **127** 1207

Landau L D and Lifshits E M 1951 *Mekhanika Sploshnykh Sred* (Moscow: Nauka) (in Russian)

Llghthill M J 1965 *J. Inst. Appl. Math.* **1** 269

Oraevsky V N 1984 *Handbook of Plasma Physics* vol 2, ed M N Rosenbluth and R Z Sagdeev (Amsterdam: North-Holland) pp 37–79

Oraevsky V N, Chodura R and Feneberg W 1968 *Plasma Phys.* **10** 819

Oraevsky V N and Pavlenko V P 1969 *Zh. Tekhn. Fiz.* **39** 1799

Oraevsky V N, Pavlenko V P, Wilhelmsson H and Kogan E Ja 1973 *Phys. Rev. Lett.* **30** 49

Oraevsky V N and Sagdeev R Z 1962 *Sov. Phys.–Tech. Phys.* **7** 955

Oraevsky V N and Tsytovich V N 1967 *Zh. Exp. Teor. Fiz.* **53** 1116

Rudakov L I and Sagdeev R Z 1958 *Fizika Plazmy i Problema Upravliaemykh Termoiadernykh Reaktsii* vol 3 (Moscow: Nauka) p 268 (in Russian)
Timofeev A V 1966 *Pisma JETP* **4** 48
Vedenov A A and Rudakov L I 1964 *Sov. Phys.–Dokl.* **9** 1073
Woodbary E J and Ng W K 1962 *Proc. IRE* **50** 2367
Zakharov V E 1966 *Zh. Exp. Teor. Fiz.* **61** 1107

non-equilibrium may be, it obeys the rule of the fastest possible relaxation to the equilibrium state. This logic is quite acceptable from the thermodynamic standpoint.

2.2 Group methods of analyzing secondary quasi-stationary states

2.2.1 Introduction

We have discussed above the evolution of a system with multiparametric non-equilibrium. It is of interest to study the properties of intermediate quasi-stationary states because the set of characteristic relaxation times may differ greatly for different secondary instabilities; among other aspects, this is necessary for the analysis of their stability. In this section we make use of symmetry arguments to analyze the properties of the turbulent quasi-stationary state. We apply the group of scaling transformations and the resulting scaling relations, and also the supersymmetry group (we follow the work of Moiseev *et al* 1988, Gendenshtein and Krive 1985, and Feigelman and Tsvelikh 1982).

Kolmogorov (1941) was the first to formulate a clear concept of turbulence scaling and the related power spectra. It was found that the idea of scaling in the turbulence of incompressible liquid enables one to determine completely the turbulence spectrum in the inertia-dominated interval because dimensional arguments are sufficient for finding critical exponents. The Kolmogorov spectrum obtained in this manner, $E(k) \sim k^{-5/3}$, was found to be in good agreement with experimental data (Monin and Yaglom 1967). It was later shown that scaling and power spectra are typical for many systems with strong and even weak interaction, for a non-equilibrium system. As examples, we can point to second-order phase transitions (Ma 1976, Patashinskii and Pokrovskii 1982), the theory of weak turbulence (Zakharov 1984), kinetic equations for particles in the presence of a source and a sink (Kats *et al* 1975, 1976) etc.

As we know, the Kolmogorov–Obukhov ideology is based on the idea that the scaling solution in a turbulent incompressible fluid corresponds to a constant energy flux in scale space (Monin and Yaglom 1967). The ideology was fully confirmed by an analysis of exactly solvable problems in the theory of weak wave turbulence (Zakharov 1965, 1984, Kats and Kontorovich 1973) which allows closed formulation in terms of the kinetic equation for waves. Among other things, it was possible to show that the collision integral possesses exact power-series scaling solutions that correspond to constant flux in the scale space of a specific integral of motion (Zakharov 1965). It became clear after this work of Zakharov that scaling is only possible if locality holds, that is, if the integrals converge in the ranges of high and low momenta, where the sink and source are. Furthermore, the results of work by Kats *et al* (1975, 1976) implied that scaling solutions also exist for kinetic equations for particles.

2.2.2 Scaling transformation group

We will mostly use the closed formulation of the strong turbulence problem in terms of the characteristic functional φ.

For an incompressible liquid we have

$$\varphi = \left\langle \exp\left(\mathrm{i} \int \boldsymbol{V}(\boldsymbol{x}, t)\boldsymbol{y}(\boldsymbol{x}, t)\,\mathrm{d}\boldsymbol{x} \right) \right\rangle \tag{2.2.1}$$

where $\boldsymbol{y}$ is an arbitrary vector field that vanishes at infinity, the velocity field $\boldsymbol{V}(\boldsymbol{x}, t)$ satisfies the Navier–Stokes equation, and averaging in (2.2.1) is carried out over the statistical ensemble. It is not difficult to obtain a linear evolution equation in variational derivatives D of the type (Hopf 1952)

$$\mathrm{i}\frac{\partial \varphi}{\partial t} = \hat{L}\varphi + \hat{I}\varphi. \tag{2.2.2}$$

The operator $\hat{L}$ is defined as follows

$$\hat{L} = -\int \mathrm{d}x^3 \boldsymbol{y} \frac{\partial}{\partial x_i} \boldsymbol{D} D_i + \mathrm{i}\nu \int \mathrm{d}x^3 \boldsymbol{y}\,\Delta \boldsymbol{D} \tag{2.2.3}$$

where ν is viscosity, $D_i = \delta/\delta y_i(\boldsymbol{x})\,\mathrm{d}^3 x$, and the linear operator $\hat{I}$ is related to the random external Gaussian force f_i which is assumed to be δ-correlated in time with the spatial correlator $B_{ij}(\boldsymbol{x}_1 - \boldsymbol{x}_2)$:

$$\hat{I} = \frac{1}{2} \int \mathrm{d}x_1^3\, \mathrm{d}x_2^3\, y_i(\boldsymbol{x}_1)\, y_i(\boldsymbol{x}_2) B_{ij}(\boldsymbol{x}_1 - \boldsymbol{x}_2) \tag{2.2.4}$$

$$\langle f_i(\boldsymbol{x}, t) f_i(\boldsymbol{x}_1, t_1) \rangle = B_{ij}(\boldsymbol{x} - \boldsymbol{x}_1)\delta(t - t_1). \tag{2.2.5}$$

It can readily be shown that the stationarity condition for the turbulence energy, which signifies the source–sink balance,

$$\langle V_i f_i \rangle = \frac{\nu}{2} \left\langle \left[\frac{\partial V_i}{\partial x_k} + \frac{\partial V_k}{\partial x_i} \right]^2 \right\rangle \equiv \bar{\varepsilon} \tag{2.2.6}$$

where $\bar{\varepsilon}$ is the mean energy dissipation rate, implies the restriction on the spatial part of (2.2.5):

$$\tfrac{1}{2} B_{ii}(0) = \bar{\varepsilon}.$$

Then the B_{ij} tensor which has, under the assumption of uniform random force, only six independent components with characteristic scales $L_m(L_1, L_2, \ldots L_6)$, can be written as

$$B_{ij}(0) = 2\bar{\varepsilon} b_{ij}.$$

Since the field $\boldsymbol{f}$ can be regarded, without loss of generality, as solenoidal, the number of independent spatial scales L_m changes to three. A more detailed structure of the tensor b_{ij} is irrelevant to this discussion.

Equations of the type of (2.2.2), (2.2.3) and (2.2.4) can also be obtained for more complex cases of compressible liquid (Moiseev *et al* 1976), magnetic hydrodynamics (Moiseev *et al* 1977), thermally stratified liquid (Sazontov 1979), or liquid in the Coriolis force field (Loginov 1980). Even though (1.2.2)-type equations cannot be solved directly, important information can be extracted from them using the scaling transformation group, and in many cases turbulence spectra can be found.

The advantages of this functional method lie in its compactness: the description of the random wave field reduces to the analysis of a single linear equation. The disadvantage is equally obvious: the variational derivatives make it quite difficult to arrive at the explicit form of solution. It was in response to this shortcoming that Kats and Kontorovich (1973) suggested a different approach described below: to attempt to analyze the mean and spectral properties of strong turbulence on the basis of the transformational properties of the equation for the characteristic functional with respect to the group of scaling transformations. Taking into account the physically obvious fact of steeply decreasing energy spectrum in the viscous range, it is also possible to extract with excellent accuracy the relation between the global characteristics of turbulence, such as the total energy and the mean energy dissipation rate.

As for the spectral properties of turbulence, we can assume the hypothesis of locality of the spectrum in the inertial subrange; we can also derive the explicit analytical expression in a number of interesting cases.

It is immediately clear that equation (2.2.2) and the initial condition

$$\varphi_{t=0} = 1 \tag{2.2.7}$$

are invariant under the group of transformations (Hopf 1952)

$$\begin{gathered} \alpha \boldsymbol{x} = \boldsymbol{x}' \qquad \alpha^{1-\beta} t = t' \qquad \alpha L_i = L_i' \quad (i = 1, 2, 3) \\ \alpha^{-(\beta+3)} \boldsymbol{y}(\boldsymbol{x}) = \boldsymbol{y}'(\boldsymbol{x}') \qquad \alpha^{1+\beta} \nu = \nu' \qquad \alpha^{3\beta-1} \bar{\varepsilon} = \bar{\varepsilon}' \end{gathered} \tag{2.2.8}$$

where α and β are arbitrary parameters. This invariance implies the similarity theorem (Hopf 1952)

$$\varphi([\boldsymbol{y}(\boldsymbol{x})]; \nu; \bar{\varepsilon}; L, t) = \varphi([\boldsymbol{y}'(\boldsymbol{x}')]; \nu'; \bar{\varepsilon}'; L', t') \tag{2.2.9}$$

which, in contrast to similarity hypotheses, is an established fact.

Rewriting the turbulence spectrum $E(k)$ in terms of φ, we find

$$E_{ij}(k) = -\frac{1}{(2\pi)^3} \int_{-\infty}^{+\infty} \mathrm{e}^{-\mathrm{i}k_0 z} \{D_i D_j' \varphi\}\Big|_{y=0} \mathrm{d}r^3, \qquad \boldsymbol{k}_0 = \frac{\boldsymbol{k}}{k}. \tag{2.2.10}$$

Using formula (2.2.9), we transform (2.2.10) to the functional that depends on the primed variables. Assuming $\alpha = k$ and integrating (2.2.10) over $k r$, we obtain for infinite integration limits

$$E_{ij}(k, t) = k^{-(2\beta+3)} f(k^{1-\beta} t, k^{1+\beta} \nu, k^{3\beta-1} \bar{\varepsilon}, kL) \Psi_{ij}(\boldsymbol{k}_0). \qquad (2.2.11)$$

In the stationary case, $\frac{\partial}{\partial t} E_{ij} = 0$. Since β is arbitrary, we find $\frac{\partial}{\partial \beta} E_{ij} = 0$. This yields the equation

$$\xi_1 \frac{\partial f}{\partial \xi_1} + \xi_2 \frac{\partial f}{\partial \xi_2} = 2f$$

$$\xi_1 = k^{1+\beta} \nu \qquad \xi_2 = k^{3\beta-1} \bar{\varepsilon}$$

whose solution can be written as

$$f = \xi_2^{2/3} f_1(\xi_1 / \xi^{1/3}).$$

Integration over angles then gives

$$E(k) = \varepsilon^{2/3} k^{-5/3} f_2(kL, kl) \qquad (2.2.12)$$

where $l = \nu^{3/4} \bar{\varepsilon}^{-1/4}$.

Formula (2.2.12) shows that the Kolmogorov spectrum follows if an additional hypothesis is made on the existence of an asymptotic expansion of the function $f(kL, kl)$ for

$$kl \ll 1, \qquad (kL)^{-1} \ll 1.$$

In fact, this assumption is equivalent to the locality hypothesis and is essentially identical to Kolmogorov's hypotheses.

The formal scaling outlined above proves to be very useful for considering more complex cases of strong turbulence in which additional parameters make themselves felt. For example, the additional parameter appearing in a compressible fluid is the velocity of sound, c_0. A direct use of dimensional arguments does not allow the calculation of the spectrum; however, the application of the appropriate invariance group for equation (2.2.2) leads one much further.

The reason for this is that the similarity group in the case of compressible liquid is found to have three parameters (Moiseev *et al* 1976). We can make use of the arbitrary parameters of the group in such a way that some of the dimensional constants are included in the initial condition to equation (2.2.2). If we now introduce a hypothesis on the complete memory loss in the initial condition, the parameters in the initial condition do not appear in the stationary spectrum. The equations for the characteristic functional in the case of compressible fluid and the similarity group were given by Moiseev *et al* (1976). We will not give here the resulting very cumbersome expressions. Note only that the assumption of complete self-similarity yields the energy spectrum for vortices (Moiseev *et al* 1976),

$$E(k) = \text{const} \cdot \rho_0 c_0^{2/(1-3\gamma)} \bar{\varepsilon}^{2\gamma/(3\gamma-1)} k^{-(5\gamma-1)/(3\gamma-1)} \tag{2.2.13}$$

where γ is related to pressure in a compressible fluid as

$$P = c_0^2 \rho_0 (\rho/\rho_0)^\gamma . \tag{2.2.14}$$

Curiously enough, spectrum (2.2.13) formally transforms to the Kolmogorov spectrum as $\gamma \to \infty$. In reality, $1 \leqslant \gamma < \infty$, so that all spectra (2.2.13) are between the Kadomtsev–Petviashvili k^{-2} spectrum (Kadomtsev and Petviashvili 1973) and the Kolmogorov $k^{-5/3}$ spectrum.

Following the scheme given above, we can obtain spectra in a thermally stratified fluid (Sazontov 1979) or fluid in a Coriolis force field (Loginov 1980).

Returning to incompressible fluids, we note that formula (2.2.12) obtained without additional hypotheses allows one to derive rigorously the so-called 'first law of turbulence' (Moiseev *et al* 1983). Integrating (2.2.12), we obtain

$$\langle E \rangle = (\bar{\varepsilon} L)^{2/3} \int_0^\infty f\left(\frac{1}{L} z; z\right) \mathrm{d}z. \tag{2.2.15}$$

In the case of $1/L \ll 1$ we find from (2.2.15) that $\langle E \rangle = (\bar{\varepsilon} L)^{2/3} \cdot \text{const}$, that is, the 'first law of turbulence' was derived here with accuracy as described above but (sic!) without any hypotheses.

Likewise, we obtain for pressure fluctuations

$$\langle p^2 \rangle - \langle p \rangle^2 = \text{const} \cdot \bar{\varepsilon}^{4/3} L^{1/3}. \tag{2.2.16}$$

This yields the universal equation of state in the theory of strong turbulence (Kadomtsev and Petviashvili 1973):

$$\langle E \rangle = \text{const} \cdot \{(\langle p^2 \rangle - \langle p \rangle^2) L\}^{1/2}. \tag{2.2.17}$$

The universality of this equation lies in the fact that deriving a formula for compressible liquids (Moiseev *et al* 1983), we arrive at the same equation (2.2.17); no stratification or rotation are observed.

2.2.3 Supersymmetry group

The previous subsection was meant to illustrate the potentials of the scaling transformation group. If we mean time-independent distributions, this group is especially efficient when it is necessary to identify self-similar stationary flux characteristics that are specific for concrete subsystems. There exist, however, non-specific completely equilibrium stationary characteristics such as the Gibbs distribution. Even if a system possesses self-similar regions of spectrum with flows, this does not exclude the existence of equilibrium subsystems. The first to point out this situation in turbulence was Millionschikov (1941) who formulated the hypothesis of the quasi-normal Gaussian distribution existing in energy-carrying regions of the spectrum. This hypothesis is physically understandable: the distribution in energy-carrying regions is only slightly disturbed by the relatively weak fluxes. The question is: how to find such areas? and what are the criteria of violation of quasi-equilibrium distributions? Supersymmetry group theory is found to be very useful here (see, for example, Gendenshtein and Krive 1985). It is not our purpose here to study the properties of supersymmetry but only to convey the minimum of information that can clarify the interest to this group in the case of turbulence.

The main property of supersymmetry is that it merges, in a very non-trivial manner, the continuous transformations (such as translations) and discrete transformations of special type (such as mirror reflection). This approach is especially popular in quantum field theory (see, for example, Volkov and Akulov 1972 and Golfand and Likhtman 1971), where the supergroup that joins the boson and fermion operators was proposed for the first time. The difference between 'matter' (fermions) and 'interactions' (bosons) was successfully eliminated only in supersymmetric theories. Divergences—the plague of quantum field theory—were also significantly reduced by this approach.

Quantum-mechanical models possessing supersymmetry properties have a number of peculiar features in the field of operator and spectral properties. For example, the Hamiltonian of such systems can be written as a Hermitian operator squared, while the ground state is non-degenerate and has zero energy. We say that supersymmetry is spontaneously broken if, with the Hamiltonian behaving 'regularly', the zero-energy ground state (the vacuum) is either absent or is destroyed by a supersymmetric transformation.

In what sense is the material above of interest to statistical physics in general and turbulence in particular? To clarify it at least at a qualitative level, we will consider a very typical example from stochastic dissipative dynamics, following Feigelman and Tsvelikh (1982):

$$\dot{\varphi}_x = -\Gamma\frac{\partial V}{\partial \varphi} + \xi(x,t) \qquad \left(\dot{\varphi}_x \equiv \frac{\partial}{\partial t}\left(\frac{\partial \varphi}{\partial x}\right)\right) \tag{2.2.18}$$

where $\xi(x,t)$ is additive noise,

$$\langle \xi(x,t)\xi(x',t')\rangle = 2T\Gamma\delta(x-x')\,\delta(t-t') \tag{2.2.19}$$

$V\{\varphi_x\}$ is the functional of the variables φ_x, and Γ is the kinetic coefficient. Equation (2.2.18) can describe both the relaxation to the thermodynamic equilibrium of a classical statistical system with energy $V\{\varphi_x\}$ and temperature T, and a broad class of relaxation processes in physical and non-physical systems far from thermodynamic equilibrium (when T is not temperature). A stochastic equivalent of (2.2.18) is the Fokker–Planck equation (FPE) for the distribution functional $\mathcal{P}\{\varphi\}$:

$$\frac{1}{\Gamma}\frac{\partial \mathcal{P}}{\partial t} = \int_{(x)} \frac{\delta}{\delta\varphi_x}\left(T\frac{\delta\mathcal{P}}{\delta\varphi_x} + \frac{\delta V}{\delta\varphi_x}\mathcal{P}\right). \tag{2.2.20}$$

Equation (2.2.20) has an obvious time-independent solution

$$\mathcal{P}\{\varphi_x\} = A\exp\left(-\frac{1}{T}V\{\varphi_x\}\right) \tag{2.2.21}$$

which yields the 'equilibrium' distribution function (in the sense that detailed balance is possible) when the normalized integral converges ($A < \infty$). The Fokker–Planck equation (2.2.20) can be converted to the equivalent Schrödinger equation whose ground state exists if $\mathcal{P}\{\varphi_x\}$ is normalizable. After Feigelman and Tsvelikh (1982) proved the supersymmetry of the fields described by this Schrödinger equation, the question of the normalizability or non-normalizability of (2.2.21) becomes equivalent to asking whether supersymmetry in the systems indicated is spontaneously broken or not. For instance, spontaneous symmetry breaking signifies that the stationary solutions of the FPE (2.2.20) are non-equilibrium (current or flow solutions).

Let us consider now some examples that illustrate the causes of supersymmetry breaking in stochastic systems (Altaisky *et al* 1990).

We begin with a Langevin system with multiplicative noise

$$\frac{\partial\varphi}{\partial t}(x,t) + \frac{\delta V[\varphi]}{\delta\varphi} - \xi(x,t)\varphi(x,t) = 0 \tag{2.2.22}$$

where ξ is the Gaussian white noise:

$$\langle\xi'(x',t')\xi(x,t)\rangle = 2\delta(x-x')\,\delta(t-t'). \tag{2.2.23}$$

Substitution of $\varphi = \exp(u)$ reduces (2.2.22) to the system with additive noise, discussed above,

$$\frac{\partial u}{\partial t}(x,t) + \mathrm{e}^{-u}V'[\mathrm{e}^{u}] - \xi(x,t) = 0 \qquad V' \equiv \frac{\delta V}{\delta(\exp u)} \tag{2.2.24}$$

which generates a supersymmetry dynamics, provided there is a functional $\Omega(u)$, such that

$$\frac{\delta\Omega(u)}{\delta u} = \mathrm{e}^{-u}V'[\mathrm{e}^{u}].$$

For the power-series potential

$$V[\varphi] = \sum_{n=0}^{\infty} a_n\varphi^n$$

we find the following Ω:

$$\Omega[u] = -a_1 \exp(-u) + 2a_2 u + \sum_{n=3}^{\infty} a_n \frac{n}{n-2} \exp(u(n-2)) + \text{const}. \tag{2.2.25}$$

We readily see that the normalized integral

$$A = \int Du \exp(-\Omega[u])$$

in a theory with potential $\Omega[u]$ may diverge and the supersymmetry may be broken even in those cases in which the equilibrium state in an appropriate Langevin system with additive noise (2.2.18) exists and is normalizable.

So far this section has dealt only with the scalar Langevin equation (2.2.18); in fact, all constructs dealing with supersymmetry are still valid in the vector case,

$$\frac{\partial \varphi_i(x,t)}{\partial t} + F_i[\varphi] - \xi_i(x,t) = 0$$

provided the potentiality conditions

$$\frac{\delta F_i}{\delta \varphi_j} = \frac{\delta F_j}{\delta \varphi_i}$$

hold; if these conditions are violated, the supersymmetry is broken and a flux arises in the system.

An example of this symmetry breaking is the process of cascade generation in the cascade model (Gledzer *et al* 1981), which is the Galerkin approximation of the initiation of turbulence. The first three equations of the cascade model are

$$\begin{aligned} \frac{\partial V_0}{\partial t} &= -\lambda_0 V_0 + f + \rho_0 (V_1^2 - V_2^2) \\ \frac{\partial V_1}{\partial t} &= -\lambda_1 V_1 - \rho_0 V_0 V_1 \\ \frac{\partial V_2}{\partial t} &= -\lambda_1 V_2 + \rho_0 V_0 V_2 \end{aligned} \tag{2.2.26}$$

where f is the Gaussian δ-correlated random force.

It can be easily shown that if $|V_0| < (\lambda_1/\rho_0)$, the second-level modes V_1 and V_2 are exponentially suppressed and system (2.2.26) is effectively reduced to a Langevin-type equation

$$\frac{\partial V_0}{\partial t} = -\lambda_0 V_0 + f.$$

In this sense, the system is supersymmetric, in accordance with Millionschikov's hypothesis. When $|V_0|$ reaches its critical value λ_1/ρ_0, second-level modes V_1 and V_2 start to grow, which results in supersymmetry breaking. Physically, this is connected with a flux arising in the system.

2.3 Effective collision frequencies in turbulent media

This section is devoted to the main ideas of the theory of turbulence in collisionless plasmas. At the same time, the material is presented in such a way that generalizations to other gaseous hydrodynamic media are clearly traceable.

We begin with the notion of collisionless plasmas. This does not mean that no interactions occur in the medium. We neglect pairwise collisions of particles which are described in kinetic equations for electrons and ions by Landau-type integrals, but retain self-consistent electromagnetic fields. This approach allows us to take into account the collective interactions of particles. The set of equations obtained is the set of kinetic equations for electrons and ions, plus Maxwell's equations for self-consistent electromagnetic fields.

In the simplest case, when we are interested in high-frequency processes in plasmas without magnetic field, we can ignore the motion of ions. Ions can be regarded as a stationary background (over one period of high-frequency motion). Then the set of equations for one-dimensional motion (along x) is greatly simplified:

$$\left.\begin{aligned} &\frac{\partial f}{\partial t} + v\frac{\partial f}{\partial x} - \frac{e}{m}E\frac{\partial f}{\partial v} = 0 \\ &\frac{\partial E}{\partial x} = 4\pi ne\left(1 - \int f\,\mathrm{d}v\right) \end{aligned}\right\} \tag{2.3.1}$$

where E is the self-consistent electric field, f is the electron distribution function, and v is the velocity of electrons along the x-axis.

For turbulent plasmas, it is convenient to write the electron distribution function as a sum of slowly and rapidly oscillating components (we follow Galeev and Sagdeev 1984):

$$f = f_0(v,t) + \delta f(v,x,t). \tag{2.3.2}$$

The slowly varying part of f_0 in (2.3.2) is so defined that averaging over rapidly varying spatial-temporal oscillations of the total distribution function f coincides with f_0, that is,

$$\langle f\rangle = f_0. \tag{2.3.3}$$

Since the function f_0 varies slowly over one period of oscillation, we can seek δf as an appropriate Fourier transform (in the WKB approximation in time). This applies to the expansion of the rapidly oscillating electric field, therefore

$$\delta f(x,v,t) = \sum f_k(v,t)\exp[-(\mathrm{i}\omega t - \mathrm{i}kx)] \tag{2.3.4}$$

$$E(x,t) = \sum E_k(t)\exp[-(\mathrm{i}\omega t - \mathrm{i}kx)] \tag{2.3.5}$$

where $\mathrm{Im}\,\omega_k = 0$.

It is also easy to write the equation for the part of the distribution function f_0 that varies slowly with time:

$$\frac{\partial f_0}{\partial t} = \left(\frac{e}{m}\right)\left\langle E\frac{\partial}{\partial v}\delta f\right\rangle. \tag{2.3.6}$$

In this derivation we made use of definition (2.3.3) and also the fact that the electric field is zero. Substituting the expansions (2.3.4) and (2.3.5) into (2.3.6), we obtain

$$\frac{\partial f_0}{\partial t} = \frac{e}{m}\sum_k E_k^*\frac{\partial}{\partial v}f_k(v,t). \tag{2.3.7}$$

The terms with $k = k_1 \neq k_2$ vanish as a result of averaging. In agreement with the linear theory of waves (Oraevsky 1984), which relates the Fourier coefficient $f_k(v,t)$ to the corresponding Fourier transform for the electric field, we have

$$f_k(v,t) = \frac{e}{m}E_k(t), \qquad \frac{\partial f_0}{\partial v}\left(P\frac{\mathrm{i}}{\omega - kv + \mathrm{i}\gamma_k} + \pi\delta(\omega_k - kv)\right) \tag{2.3.8}$$

where the symbol P indicates that we have excluded the range of electron velocities for which $|\omega_k - kv| < \gamma_k$. It is customary, following Vedenov *et al* (1961), to refer to equation (2.3.7) together with (2.3.8) as the quasi-linear equation for the velocity distribution function f_0. It takes the form of the diffusion equation in the velocity space:

$$\frac{\partial f_0}{\partial v} = \frac{\partial}{\partial v}D\frac{\partial f_0}{\partial v}. \tag{2.3.9}$$

The diffusion coefficient $D(v)$ in (2.3.9) is defined (as implied by (2.3.8)) by

$$D(v,t) = \frac{e^2}{m^2}\sum |E_k|^2\left(P\frac{\gamma_k}{(\omega - kv)^2 + \gamma_k^2} + \pi\delta(\omega_k - kv)\right). \tag{2.3.10}$$

This equation must be complemented with an equation for the evolution of oscillation amplitude E_k, and also with an equation for the dependence of growth rates (or decay rates) on the form of the distribution function f_0. These equations can also be written on the basis of the results of the linear theory (Vedenov *et al* 1961), taking into account the WKB approximation used,

$$\frac{\partial}{\partial t}|E_k(t)|^2 = 2\gamma_k|E_k(t)| \tag{2.3.11}$$

where

$$\gamma_k(t) = \frac{\pi}{2}\omega_k\left(\frac{\omega_{pe}}{k}\right)\frac{\partial f_0(v,t)}{\partial v}\bigg|_{\omega=kv}. \tag{2.3.12}$$

Equations (2.3.9)–(2.3.12) give a complete description of plasma behavior in the quasi-linear approximation.

As shown by (2.3.9), we now have an equation that formally contains a collision integral (the right-hand side). This collision integral has, on the one hand, the form of the diffusion integral in the velocity space (similarly to the integral of pairwise collisions), and on the other hand, it is determined by the oscillation spectrum $|E_k|^2$. The frequencies of energy–momentum transfer calculated using (2.3.9) are known as effective collision frequencies. Without boring the reader with lengthy manipulations, we just mention that the general expression for the quasi-linear equation that takes into account arbitrary waves in magnetoactive plasma was derived by Vedenov *et al* (1961). Note also that the quasi-linear approximation in inhomogeneous magnetoactive plasma (Galeev and Rudakov 1963) leads not only to diffusion in the velocity space but also to diffusion terms in ordinary space. Moiseev and Sagdeev (1963) were the first to prove that instabilities due to drift waves lead to diffusion in ordinary space.

2.4 Induced drift instabilities

This is a familiar fact that taking into account dissipative effects may result in plasma instabilities (Kogan *et al* 1965). For example, the electron–ion friction generates drift–dissipative instability, which can give a qualitative explanation to Bohm's diffusion (Moiseev and Sagdeev 1963). Nevertheless, these instabilities, arising due to collisions, automatically fade out and ultimately vanish completely in the transition to a collisionless regime (including the high-temperature regime). Still, the very fact of the existence of non-equilibrium makes it possible for induced instabilities to arise.

In this section we consider the effect of high-frequency turbulent fluctuations on the evolution of a special type of drift–dissipative instability. This instability is caused by the appearance of electron viscosity. We show in the next section that the high-frequency turbulence results in effective 'collision' integrals. We need to emphasize that, as follows from the analysis in the preceding section, these 'collision' integrals do not affect the ions' motion. It is not difficult to show that the electron component conserves its momentum. This leads to an important conclusion that electron–ion friction effects are absent, while the predominant dissipative process is the electron viscosity. It can readily be shown that the electron viscosity in magnetoactive inhomogeneous plasma results in drift-wave instability. To show it, consider the simplest case: one-dimensional plasma inhomogeneity along the x-axis, the z-axis pointing along the constant magnetic field (see also section 2.1). Note that this instability is similar to the instability dictated by the electron–ion friction (Galeev *et al* 1963). Indeed, an additional term $\eta\frac{\partial^2 v}{\partial z^2}$ arises in the case of electron viscosity in the equation for the longitudinal electron motion (along the magnetic field $\boldsymbol{H}_0$); this term is analogous, in the long run, to the term $mn_x\nu_{ei}v_z$. Therefore, if $\nu_{ei}\to 0$ and the electron friction coefficient is finite, the drift instability due to electron friction is generated. To prove this, it is sufficient to write the equation of motion for ions (assuming for simplicity cold ions) moving across the magnetic field,

$$MN_{0i}\frac{\partial V_i}{\partial t} = eN_{0i}\boldsymbol{E}_{\perp} + \frac{e}{c}N_{0i}[V_i \times \boldsymbol{H}_{0i}] \tag{2.4.1}$$

and also the equation for the longitudinal motion of electrons, in which the inertia of electrons is omitted, as is typical for drift waves:

$$-\mathrm{i}k_z \delta N_e T_0 - eN_{0e}E_z - k_z^2 \tau N_{0e} T_0 V_e = 0. \tag{2.4.2}$$

This system must be complemented with the condition of quasi-neutrality of drift motion, that is,

$$\operatorname{div} \boldsymbol{j}_e = 0 \tag{2.4.3}$$

where $\boldsymbol{j}$ is the total electric current.

The set of equations (2.4.1)–(2.4.3) immediately gives the dispersion equation

$$1 - \mathrm{i}\frac{\omega_s}{\omega k_z^2 \lambda_e^2}\left[1 - \frac{\omega_e}{\omega}\right] = 0 \tag{2.4.4}$$

where

$$\omega_s = \frac{\Omega_i \Omega_e}{\nu_e}\frac{k_z^2}{k_y^2} \qquad \omega_e = \frac{N'_{0e}}{N_{0e}}k_y\frac{cT}{eH}. \tag{2.4.5}$$

Here N_e and N_i are the perturbed electron and ion densities, respectively, in the plasma, T is the temperature, Ω_i and Ω_e are the ion and electron cyclotron frequencies, ν_e is the effective frequency of electron collisions, λ_e is the appropriate free path length, and N'_{0e} is the derivative of density with respect to x. In the case of $\omega_s^* \gg \omega_e$, this equation has the solution

$$\operatorname{Im}\omega \sim \frac{\omega_e^2 k_z^2 \lambda_e^2}{\omega_s} \qquad \omega_s^* = \frac{\omega_s}{k_z^2 \lambda_e^2}. \tag{2.4.6}$$

Note first of all that when conventional hydrodynamics is applicable, the viscous term due to pairwise collisions is less by a factor of $(\lambda_e/\lambda_{ei})^2$ than the term due to the electron friction. Therefore, our example with the viscous term but with zero electron–ion friction holds for plasmas with high-frequency turbulence.

The most realistic is the HF turbulence produced by a beam of charged particles. The quasi-linear equation derived in the preceding section can in this case be recast in the form

$$\frac{\partial f_0}{\partial t} = \frac{e^2}{m^2}\frac{\partial}{\partial v}\sum\left(\frac{E_k^2}{v}\frac{\partial f_0}{\partial v}\right) \tag{2.4.7}$$

where f_0 is in fact the beam distribution function, since this is the range of velocities in which particles strongly interact with the waves generated. In order to evaluate the effect of high-frequency oscillations on the drift–dissipative instability of interest to us now, it is in fact necessary to find the effective time of collisions for beam electrons and substitute it into (2.4.6). One has to remark first, though, that an equation of the type of (2.4.2) should be written not for all N_e electrons of the plasma but only for beam electrons, since only these electrons acquire viscosity. This means that instead of (2.4.2), we have

$$-ik_z\delta n_e T_0 - en_{0e}E_z - k_z^2\tau n_{0e}T_0V_e = 0 \qquad (2.4.2a)$$

with equation (2.4.1) still holding. Obviously, (2.4.6) is replaced in this case by

$$\operatorname{Im}\omega \sim \frac{\omega_e^2 k_z^2 \lambda_e^2 N_{0e}}{\omega_s n_0}, \qquad \omega_s^* = \frac{\omega_s n_0}{k_z^2\lambda_e^2 N_{0e}}. \qquad (2.4.6a)$$

We now only need to find the effective time of electron collisions. This procedure was constructed by Galeev and Sagdeev (1984). Following their work, we can evaluate this time from a set of quasi-linear equations, using (2.4.7),

$$\tau_e \sim \sqrt{\frac{N_{0e}}{n_0}}\frac{1}{\omega_0^*} \qquad (2.4.8)$$

where ω_0^* is the plasma frequency for the beam density. Strictly speaking, expression (2.4.8) is valid if beam particle velocities do not appreciably exceed thermal velocities.

We have thus shown in the example of a plasma with two-parameter non-equilibrium (electron beam + inhomogeneity magnetoactive plasma) that high-frequency instability ultimately produces low-frequency instability, that is, it results in an induced instability (of drift type in this case).

References

Akhiezer A I and Fainberg Ya B 1949 *Dokl. Akad. Nauk SSSR* **69** 555

Altaisky M V, Moiseev S S and Pavlik S I 1990 *Phys. Lett.* **147** A 142–6

Baikov I S 1965 *Yadernyi Sintez* **5** 326

Bohm D and Gross E P 1949 *Phys. Rev.* **75** 1851

Chew G, Goldberger M and Low F 1956 *Proc. R. Soc.* **236** 112

Feigelman M V and Tsvelikh A M 1982 *Zh. Exp. Teor. Fiz.* **83** 1430–3

Galeev A A, Moiseev S S and Sagdeev R Z 1963 *Atomn. Energia* **15** 451–67

Galeev A A, Oraevsky V N and Sagdeev R Z 1963 *Zh. Exp. Teor. Fiz.* **44** 902

Galeev A A and Rudakov L I 1963 *Zh. Exp. Teor. Fiz.* **45** 647

Galeev A A and Sagdeev R Z 1984 *Handbook of Plasma Physics* vol 2, ed M N Rosenbluth and R Z Sagdeev (Amsterdam: North-Holland) pp 683–98

Gaponov A V and Rabinovich M I 1984 *Fizika XX Veka (Physics of the XX Century)* ed E P Velikhov (Moscow: Nauka) pp 219–80 (in Russian)
Gendenshtein L E and Krive I V 1985 *Usp. Fiz. Nauk* **146** 553–88
Gledzer E B, Dolzhansky F V and Obukhov A M 1981 *Sistemy Gidrodinamicheskogo Tipa* (Moscow: Nauka) pp 1–366 (in Russian)
Golfand Yu A and Likhtman E P 1971 *Pisma JETP* **13** 452–5
Hopf E 1952 *J. Rat. Mech. Anal.* **1** 87–123
Kadomtsev B B and Petviashvili V I 1973 *Dokl. Akad. Nauk SSSR* **208** 794–6
Kats A V and Kontorovich V M 1973 *Zh. Exp. Teor. Fiz.* **64** 153–63
Kats A V, Kontorovich V M, Moiseev S S and Novikov B E 1975 *Pisma JETP* **21** 13–6
Kats A V, Kontorovich V M, Moiseev S S and Novikov B E 1976 *Zh. Exp. Teor. Fiz.* **7** 177–92
Kogan E Ya, Moiseev S S and Oraevsky V N 1965 *Zh. Prikl. Mech. Tech. Fiz.* **6** 41–6
Kolesnichenko Ya I and Oraevsky V N 1967 *Atomn. Energia* **23** 289
Kolmogorov A N 1941 *Dokl. Akad. Nauk SSSR* **30** 299–302
Loginov V M 1980 *Zeitschrift für Phys.* B **37** 363
Ma S 1976 *Modern Theory of Critical Phenomena* (London: Benjamin)
Mikhailovsky A B 1971 *Teoriya Plazmennykh Neustoichivostei* vol 1, 2 (Moscow: Atomizdat) (in Russian)
Millionschikov M D 1941 *Izv. Akad. Nauk SSSR Ser. Fiz.* **5** 433–46
Moiseev S S and Sagdeev R Z 1963 *Zh. Exp. Teor. Fiz.* **44** 763–5
Moiseev S S, Tur A V and Janovsky V V 1976 *Zh. Exp. Teor. Fiz.* **71** 1062–73
Moiseev S S, Tur A V and Janovsky V V 1977 *Zh. Tekhn. Fiz.* **47** 473–81
Moiseev S S, Tur A V and Janovsky V V 1983 *Dokl. Akad. Nauk SSSR* **269** 600–3
Moiseev S S, Tur A V and Janovsky V V 1988 *Renormalization Group* (Singapore: World Scientific) pp 131–45
Monin A S and Yaglom A M 1967 *Statisticheskaya Gidromekhanika* Part 2 (Moscow: Nauka) (in Russian)
Oraevsky V N 1984 *Handbook of Plasma Physics* vol 1, ed M N Rosenbluth and R Z Sagdeev (Amsterdam: North-Holland) pp 451–67
Oraevsky V N, Chodura R and Feneberg W 1968 *Plasma Phys.* **10** 819
Patashinskii A Z and Pokrovskii V L 1982 *Fluktuatsionnaya Teoria Fazovykh Perekhodov* (Moscow: Nauka) pp 1–38 (in Russian)
Rudakov L I and Sagdeev R Z 1958 *Fizika Plazmy i Problema Upravliayemykh Termoiyadernykh Reaktsii* vol 3 (Moscow: Nauka) p 268 (in Russian)
Rudakov L I and Sagdeev R Z 1961 *Dokl. Akad. Nauk SSSR* **138** 581
Sazontov A G 1979 *Izv. Akad. Nauk SSSR FAO* **25** 820–8
Vedenov A A, Velikhov E P and Sagdeev R Z 1961 *Yadernyi Sintez* **1** 82
Volkov D V and Akulov V P 1972 *Pisma v JETP* **16** 621–4
Zakharov V E 1965 *Zh. Prikl. Mekh. Tekhn. Fiz.* **4** 35
Zakharov V E 1984 *Handbook of Plasma Physics* vol 2, ed M N Rosenbluth and R Z Sagdeev (Amsterdam: North-Holland) pp 3–36

Chapter 3

External noise and stability of plasmas

3.1 Introduction

We can consider as secondary not only those instabilities that were described in the previous chapters but also those that evolve in response to external noise, as a result of the resonance interaction between certain harmonics of the noise spectrum and the natural vibrations of a dynamic system. Generalizations of this type appear to be justifiable because perturbations arising in a system as a result of the initial instability can be treated as external noise with respect to perturbations of a different type that evolve in the process of secondary instability. The primary instability need not evolve to a well-developed stage because the noise level sharply increases (by two to three orders of magnitude) already at its threshold value.

It is important to note that we will consider the effects of external, not internal, fluctuations. This assumes, among other things, a source of fluctuations independent of the system given; for simplicity, this source is assumed to be fixed in advance. Thermal fluctuations (Shafranov 1963) become important and must be taken into account when the spectral maximum of thermal distribution lies close to the frequency of plasma vibrations.

We pay maximum attention to studying the effect of external multiplicative noise, that is, of random oscillations (due to the energy of the external medium) of the parameters of the system that enter as multipliers to quantities that characterize the state of the system. This choice is certainly not arbitrary. It is a known fact (Horsthemke and Lefever 1984) that the external noise (fluctuations of the medium) influences large systems much more strongly than internal fluctuations. Also, in contrast to the additive noise which can only spread out the stationary state of the system, the multiplicative noise can change it drastically. For example, fluctuations of the medium produce a qualitative modification in the evolution of non-equilibrium systems (Haken 1978), generate kinetic transitions in non-equilibrium systems with decay, reproduction and diffusion (Mikhailov and Uporov 1984), cause stochastic parametric resonance (Klyatskin 1975), and accelerate the evolution of unstable systems (Butz *et al* 1989) etc.

On the other hand, the role of internal fluctuations in system stability (the less important factor, as explained above) can only be correctly analyzed in a self-consistent manner. This is a considerably more complex problem that will be the subject of further investigation. It is already clear from thermodynamic arguments, however, that internal thermal fluctuations cannot destabilize equilibrium distributions; the interesting aspect is their effect on non-equilibrium distributions. We need to emphasize one important feature of the analysis to be given: the effect of the fixed external noise taken into account here depends on the state of the system. This is its essential distinction against the traditional analysis using additive noise, where fluctuations are used only as suppliers of the required type of perturbations.

In reality they play a much more important role; in a nutshell, the behavior of even very small fluctuations is very aggressive and thus they cannot be discarded. For example, concepts treating the development of chaos in dynamic systems undergo qualitative modification. This conclusion follows from an analysis of the simplest nonlinear dynamic system—the Duffing oscillator (Barts *et al* 1988). Indeed, the results of this paper imply an accelerated mode of instability evolution when the additive noise is taken into account. The second moments of the random field then grow faster than the first moments, that is, the system's dispersion and chaos (the so-called fluctuational chaos) both grow with time. Thus section 3.2 shows for the parametric instability that the fluctuational noise can arise at amplitudes that are much smaller than those responsible for the dynamic chaos; this means that fluctuations generate new critical states. The role of multiplicative noise (for example, of eigenfrequency fluctuations) is much greater than that of additive noise. The chaos domain is thus widened considerably by small-amplitude fluctuations. Note also that chaotization of this type is also possible in noiseless integrable systems, in which dynamic chaos is in principle impossible. No less important is the problem of instability evolution, and generally of fluctuations accelerating the passage of the system through the bifurcation point. From the thermodynamical point of view, it is obvious that evolution inevitably pushes the system to the best equilibrium state possible under given conditions. It is shown in section 3.3 that the loss of stability in response to multiplicative fluctuations, at least in the case of an isolated point, is accelerated exponentially and the system immediately 'rushes' towards a chaotic state; in general, this chaos may be intermittent. This last remark also means that the energy of growing perturbations gets concentrated with time in narrow regions of space.

Section 3.4 gives another example of destabilization of the dynamic system by the external multiplicative noise. We consider the effect of field strength fluctuations of a uniform magnetic field in which a charged particle moves, on the stability of its motion in the presence of an electromagnetic wave traveling along the magnetic field. It is shown that such fluctuations may accelerate the chaotization of particle motion and enhance the diffusion of particles in energy. Finally, section 3.5 discusses the effect of fluctuations of the viscosity coefficient on the diffusion of passive scalar or vector admixture.

3.2 Fluctuational chaos

Two problems in the physics of nonlinear media (including nonlinear physics of plasmas) were especially intriguing for researchers in the last two or three decades: dynamic chaos and self-organization (see e.g. Zaslavskii 1984 and Haken 1978). As for the former problem, we have to accept that to derive noise in a system without noise would be elegant and in fact important. However, a system inevitably has some low-level noise and neglecting it is actually based on assuming its 'non-aggressive' behavior. In what follows, we follow Butz and Moiseev (1990) and show that this assumption is not always true.

Consider the decay of a transverse wave of frequency ω to a transverse wave and a Langmuir wave ($t \to t' + l$). The initial set of equations that defines slowly varying variables is

$$\begin{gathered}\operatorname{rot}\operatorname{rot}\boldsymbol{E} - k_0^2\varepsilon\boldsymbol{E} - \frac{2\mathrm{i}k_0}{c}\frac{\partial\boldsymbol{E}}{\partial t} = -\frac{\omega_p^2}{c^2}\frac{\delta n}{n_0}\boldsymbol{E} \\ \frac{\partial^2\delta n}{\partial t^2} + \omega_p^2\delta n - v_{T_e}^2\Delta\delta n = \frac{\omega_p^2}{\omega^2}\frac{\Delta|\boldsymbol{E}|^2}{16\pi m}.\end{gathered} \tag{3.2.1}$$

where $\boldsymbol{E}$ is the envelope of the high-frequency electromagnetic field,

$$\boldsymbol{E} = \tfrac{1}{2}\left[\boldsymbol{E}\mathrm{e}^{-\mathrm{i}\omega t} + \text{c.c.}\right]$$

δn is the slowly varying component of density of plasma electrons, $k_0 = \omega/c$, $\varepsilon = 1 - (\omega_p^2/\omega^2)$, $v_{T_e}^2 = T_e/m$ (T_e is the electron temperature, m is the electron mass). We seek the solution to (3.2.1) in the form

$$\begin{gathered}\boldsymbol{E} = \boldsymbol{A}_i(t)\mathrm{e}^{\mathrm{i}\boldsymbol{k}_i\boldsymbol{r} - \mathrm{i}\delta t} + \boldsymbol{A}_s(t)\mathrm{e}^{\mathrm{i}\boldsymbol{k}_s\boldsymbol{r}} \\ \delta n = \tfrac{1}{2}\delta n_p(t)\mathrm{e}^{\mathrm{i}\boldsymbol{k}_p\boldsymbol{r}} + \text{c.c.}\end{gathered} \tag{3.2.2}$$

Note that $\boldsymbol{k}_i = \boldsymbol{k}_s + \boldsymbol{k}_p$, $\delta = \omega_i - \omega_s$ ($|\delta|/\omega_{i,s} \ll 1$). We also assume that the plasma density $n = n_0 + \tilde{n}(t)$, where $\tilde{n}(t)$ is the fluctuation component, which we assume, for the sake of simplicity, to be spatially uniform. Substituting then (3.2.2) into (3.2.1) and changing to dimensionless variables we arrive at the following set of equations for the amplitudes of the coupled oscillations:

$$\begin{aligned}\frac{\mathrm{d}C_0}{\mathrm{d}\tau} &= -\mathrm{i}\rho(1+z)\mathrm{e}^{\mathrm{i}\Delta\tau}C_1 \\ \frac{\mathrm{d}C_1}{\mathrm{d}\tau} &= -\mathrm{i}\rho^*(1+z)\mathrm{e}^{-\mathrm{i}\Delta\tau}C_0 \\ \frac{\mathrm{d}^2\rho}{\mathrm{d}\tau^2} &+ \Gamma^2(1+z)\rho = -C_0C_1^*(1+z)\mathrm{e}^{-\mathrm{i}\Delta\tau}.\end{aligned} \tag{3.2.3}$$

The following notation was introduced above:

$$C_0 = \frac{A_i}{A_0} \qquad C_1 = \frac{A_s}{A_0} \qquad \rho = \left(\frac{\mu^2}{\alpha A_0^2}\right)^{1/3} \frac{\delta n}{n_0}$$

$\boldsymbol{A}_{i,s} = \boldsymbol{e}_{i,s} A_{i,s}$, $\boldsymbol{e}_{i,s}$ are the unit polarization vectors of the electric field of the electromagnetic wave ($\boldsymbol{e}_{i,s} \boldsymbol{k}_{i,s} = 0$), $A_0 = A_{i(t=0)}$ are the initial values of the electromagnetic pump wave, $\mu \equiv \omega_p^2 \cos\theta / 4\omega_i$, $\alpha \equiv k_p^2 \omega_p^2 \cos\theta / \omega_i^2 8\pi n_0 m$, θ is the angle between the vectors $\boldsymbol{e}_i$ and $\boldsymbol{e}_s$, $\tau = (\mu\alpha A_0^2)^{1/3} t$

$$\Delta = \frac{\delta}{(\mu\alpha A_0^2)^{1/3}} \qquad z = \frac{\tilde{n}}{n} \qquad \Gamma^2 = \frac{\omega_p^2}{(\mu\alpha A_0^2)^{2/3}} \equiv \left(\frac{\omega_\rho^3 \cos^2\theta A_0^2}{32\pi\omega_i^3 n_0 m v_\phi^2}\right)^{-2/3}$$

where $v_\phi = \omega_p / k_p$.

Note that the set (3.2.3) is sufficiently universal and describes a broad class of decay processes, including the decay of a plasma wave to a plasma and an ion-acoustic wave, and also decay processes in plasma–beam systems. Only two parameters, Γ (which is inversely proportional to the pumping wave amplitude) and Δ, change here.

We will briefly summarize the results in several limiting cases. If $z = 0$ and $C_0(\tau) = \text{const} = 1$, we seek the solution of (3.2.3) in the form $\rho = \rho_0 \exp(\mathrm{i}\Omega\tau)$, $C_1 = C_{10} \exp(\mathrm{i}\Omega\tau - \mathrm{i}\Delta\tau)$. The linear stage of parametric instability is then characterized by the growth rate which reaches a maximum for $\Delta = \Gamma$ if $\Gamma \gg 1$ ($\operatorname{Im}\Omega = 1/\sqrt{2\Gamma}$), and for $\Gamma < 1$ the maximum is reached at $\Delta = 0$ ($\operatorname{Im}\Omega = \sqrt{3}/2$). In this last case a modified process develops, in which both these waves (the forward and the backward) are in resonance with the electromagnetic waves.

A numerical analysis of the nonlinear stage of parametric instability ($c_0 \neq$ const) was performed for $z = 0$ at various values of Γ and Δ. It was shown that in the case of $\Gamma \gg 1$ the coupled waves exchange energy regularly and periodically. The characteristic time period of this process is given by the inverse of the growth rate at the linear stage. If $\Gamma < 1$, the decay dynamics changes dramatically. It becomes chaotic, corresponding to overlapping resonances, $\omega_i = \omega_s \pm \omega_p$. We can now take into account small density fluctuations and investigate the case of small fixed amplitudes of the pump wave ($C_0 = 1$ and $\Gamma \gg 1$), when electromagnetic waves interact with only one plasma wave. Set (3.2.3) can be further simplified by the replacement

$$\rho(\tau) = P(\tau)\mathrm{e}^{-\mathrm{i}\Gamma\tau} \tag{3.2.4}$$

where $P(\tau)$ is a new slow variable.

Substituting (3.2.4) into (3.2.3) and assuming that the synchronization conditions hold ($\Gamma = \Delta$), we obtain a simple set of equations for the amplitudes C_1 and P:

$$\frac{\mathrm{d}P}{\mathrm{d}\tau} + \frac{1}{2}\Gamma_z P = -\frac{\mathrm{i}}{2\Gamma}(1+z)C \qquad \frac{\mathrm{d}C}{\mathrm{d}\tau} = \mathrm{i}(1+z)P \tag{3.2.5}$$

where $C = C_1^*$. Using the Furutsu–Novikov formula (see, for example, Klyatskin 1975) and also expressions for variational derivatives that follow from equations (3.2.5),

$$\begin{aligned} \frac{\delta P(\tau)}{\delta z(\tau)} &= -\frac{\mathrm{i}}{4}\left[\frac{1}{\Gamma}C + \Gamma P\right] \\ \frac{\delta C(\tau)}{\delta z(\tau)} &= \frac{\mathrm{i}}{2}P, \end{aligned} \tag{3.2.6}$$

we obtain the following equations for first and second moments:

$$\begin{aligned} \frac{\mathrm{d}\langle P\rangle}{\mathrm{d}\tau} + \frac{\mathrm{i}}{2}\Gamma D\left\{-\frac{\mathrm{i}}{2}\Gamma\langle P\rangle - \frac{\mathrm{i}}{2\Gamma}\langle C\rangle\right\} &= -\frac{\mathrm{i}}{2\Gamma}\langle C\rangle + \frac{1}{2\Gamma}D\langle P\rangle \\ \frac{\mathrm{d}\langle C\rangle}{\mathrm{d}\tau} = \mathrm{i}\langle P\rangle + \frac{D}{2}\left\{\Gamma\langle P\rangle + \frac{1}{\Gamma}\langle C\rangle\right\} \end{aligned} \tag{3.2.7}$$

$$\begin{aligned} &\frac{\mathrm{d}}{\mathrm{d}\tau}\langle P^2\rangle + D\Gamma^2\langle P^2\rangle = -\frac{\mathrm{i}}{\Gamma}\langle CP\rangle + D\left(\frac{\langle P^2\rangle}{\Gamma} - \frac{\langle C^2\rangle}{2\Gamma^2} - \langle CP\rangle\right) \\ &\frac{\mathrm{d}}{\mathrm{d}\tau}\langle CP\rangle + \frac{D\Gamma^2}{4}\langle CP\rangle - \frac{D\Gamma}{2}\langle P^2\rangle = -\frac{\mathrm{i}}{2\Gamma}\langle C^2\rangle + \mathrm{i}\langle P^2\rangle + \frac{2}{\Gamma}\langle CP\rangle - \frac{D}{4}\langle C^2\rangle \\ &\frac{1}{2}\frac{\mathrm{d}}{\mathrm{d}\tau}\langle C^2\rangle = \langle CP\rangle\left(\mathrm{i} + \frac{D\Gamma}{2}\right) + \frac{D}{2\Gamma}\langle C^2\rangle - D\langle P^2\rangle. \end{aligned} \tag{3.2.8}$$

In deriving the sets (3.2.7) and (3.2.8), we assume for simplicity that the function $z(\tau)$ is a Gaussian zero-mean δ-correlated random process

$$\langle z(\tau)\rangle = 0 \qquad \langle z(\tau)z(\tau_1)\rangle = 2D\delta(\tau - \tau_1). \tag{3.2.9}$$

An analysis of the set (3.2.7) and (3.2.8) gives the following results. If $D\Gamma^2 \ll 1$, the effect of fluctuations is small and the dynamics is still regular. If $D\Gamma^2 > 1$, first moments grow as

$$\operatorname{Im}\omega^{(1)} = \sqrt{\frac{4}{5}\frac{2}{D\Gamma^3}}\frac{\omega_p}{\Gamma}.$$

As for the second moments, we need to emphasize that if $D\Gamma^2 \gg 1$, the forward and backward plasma waves strongly interact and this mostly affects the behavior of second moments. Taking this factor into account in the equations for second moments gives rise to the expression $\operatorname{Im}\omega^{(2)} \sim D\Gamma^2(\omega_p/\Gamma) \gg \operatorname{Im}\omega^{(1)}$. As a result, dispersion grows exponentially and the system rapidly moves into the chaotic mode. In a certain sense, this growing fluctuational chaos is the 'pure' chaos because it is free of the difficulties of the dynamic chaos due to de-chaotization 'gaps' and stability 'islands'. The most important factor is, though, that the fluctuational chaos develops at significantly lower pump wave amplitudes than the dynamic chaos.

3.3 Universal noise-induced plasma instability and accelerated fluctuational dynamics in the vicinity of a point of instability

In this section we concentrate on one important peculiarity of the dynamics of a process destabilized by fluctuations at the stability threshold: accelerated evolution of the system. The practical importance of this aspect for the plasma is evident in the following example of a linear system:

$$\frac{d^2\xi}{dt^2} + \Lambda[1 + q(t)]\xi = 0. \tag{3.3.1}$$

Here $q(t)$ is a random Gaussian delta-correlated function ($\langle q(t)q(t')\rangle = D\delta(t - t')$). If $q = 0$, an equation of the type (3.3.1) describes the stability of plasma equilibrium in the ideal magnetic hydrodynamics approximation, Langmuir oscillations in cold plasma, and a number of other phenomena. If $\Lambda < 0$, (3.3.1) describes an unstable system with $\mathrm{Im}\,\omega = \sqrt{|\Lambda|}$. As can readily be seen in this case, taking $q(t)$ into account does not influence the rate of increase of the first moment $\langle \xi \rangle$, while the increment for the second moment gains a positive additional term ($\Delta\,\mathrm{Im}\,\omega = \frac{1}{4}D|\Lambda|$), that is, it grows faster. We need to stress also that taking into account the random Gaussian additional term in equation (3.3.1) leads to universal instability (also for the plasma in the MHD approximation) regardless of the sign of the parameter Λ.

We will consider this aspect in more detail, following Butz *et al* (1990). The classical analysis of plasma stability problems assumes that the external influence due to fluctuations is small and is usually neglected. External noise is known to affect all physical systems and in many cases may substantially change the course of a process. The degree to which external noise influences a specific physical system is not always immediately clear. Zaslavskii and Moiseev (1966) reduced the problem of the effect of fluctuations of the curvature of the effective gravity field in the case of the interchange (flute) perturbations of the plasma to solving the Schrödinger equation with random potential for its spectrum; only spatial fluctuations were considered. In what follows we analyze plasma behavior in the MHD approximation assuming that the statistical characteristics of noise are known. We will show that taking into account even low-amplitude multiplicative noise results in universal instability. The second example analyzed below shows that instabilities of similar type are possible for plasmas with dissipation. The cause of this instability is the parametric destabilization of the dynamic system owing to the presence of natural harmonics of the system in the noise spectrum.

Indeed, we obtain for the frequency of natural vibrations of the plasma, following the energy principle (Bernstein *et al* 1958) and assuming the perturbations of the ground state in the form $\exp(i\omega t)$,

$$\omega^2 = \frac{\int \boldsymbol{\xi}\hat{K}\boldsymbol{\xi}\,\mathrm{d}\boldsymbol{r}}{\int \rho\xi^2\,\mathrm{d}\boldsymbol{r}}$$

where $\boldsymbol{\xi}$ is a small displacement of the plasma from the equilibrium position,

$\hat{K}$ is the differential operator acting on ξ as a function of coordinates and representing the elasticity of the plasma, and ρ is the plasma density.

Let us clarify the physical mechanism of destabilizing natural oscillations of the plasma. The fluctuations of the parameters in the operator $\hat{K}$ due to external energy sources are equivalent to additional random force $\hat{K}_c\xi$ in the linearized equation of motion of the plasma, written in the form $\rho\partial^2\xi/\partial t^2 = -\hat{K}\xi - \hat{K}_c\xi$. As a visually clear model, easily illustrating the required properties and amplitude of the noise leading to destabilization of a system, we can choose a concrete oscillatory model (3.3.1) at $\Lambda = \omega^2 > 0$. This shows that the requirement for the frequency spectrum $q(t)$ in the case of the parametric resonance must be that it must spread over at least a frequency band in the vicinity of the value 2ω, so as to cover the area of maximum growth rate in parametric resonance. Details of this problem are discussed below.

We can now estimate the threshold level of the external noise at which its role is more important than that of the thermal noise. Ever since people started to use Langevin's equation, we know a way of taking thermal fluctuations into account: by introducing an additive external force whose correlation function is proportional to the system's temperature. For example, in the case of a Brownian-motion-type model it is convenient to make use of the preceding equation but add to it a linear friction, that is a simple nonlinear term and an additive external force:

$$\frac{d^2\xi}{dt^2} + \nu\frac{d\xi}{dt} + \omega^2(1 + q(t))\xi + \alpha\xi^2 = f(t). \tag{3.3.2}$$

Here the system is assumed to be δ-correlated (the correlator $\langle f(t)f(t+\tau)\rangle = \nu T\delta(\tau)$, T being the temperature of the system). Recalling that the nonlinear term is neglected when analyzing a system's stability, we can easily obtain from equation (3.3.2) the criterion of smallness of the effect of thermal fluctuations:

$$\sqrt{D} > \alpha\sqrt{\nu T}/\omega^4. \tag{3.3.3}$$

Obviously, the (3.3.3)-type criterion changes somewhat for each specific problem. However, the threshold value D characterizing the correlation properties of the multiplicative noise will rise with increasing temperature of the system (e.g. plasma temperature).

If a (3.3.3)-type condition is satisfied, this means that we study the stability of the plasma subjected to perturbations that are small—compared with the steady-state values—but finite.

One important peculiarity in the behavior of a plasma with fluctuating parameters must be mentioned. As will be shown below, the presence of random oscillations may result in intermittency. Therefore, anomalous diffusion can also be intermittent.

To illustrate the situation, consider two examples in which a small stochastic perturbation added to an ordinary dynamic system leads to plasma instability of the type described above.

The linearized equation of motion for small displacements of the ideal plasma from the equilibrium position has the form (Kadomtsev 1963)

$$\rho_0 \frac{\partial^2 \boldsymbol{\xi}}{\partial t^2} = \nabla(\gamma P_0 \operatorname{div} \boldsymbol{\xi} + \boldsymbol{\xi} \nabla P_0) + \frac{1}{4\pi}[\operatorname{rot} \operatorname{rot} [\boldsymbol{\xi}, \boldsymbol{B}_0] \boldsymbol{B}_0] + \frac{1}{4\pi}[\operatorname{rot} \boldsymbol{B}_0, \operatorname{rot} [\boldsymbol{\xi}, \boldsymbol{B}_0]] \tag{3.3.4}$$

where ρ_0, $\boldsymbol{B}_0$ and P_0 are the non-perturbed equilibrium values of density, magnetic field, and pressure, and γ is the adiabatic exponent.

This equation can be rewritten as

$$\rho_0 \frac{\partial^2 \boldsymbol{\xi}}{\partial t^2} = -\hat{K} \boldsymbol{\xi} \tag{3.3.5}$$

where $\hat{\boldsymbol{K}}$ is a linear self-conjugate operator.

Let us consider small random perturbations of the magnetic field $\boldsymbol{B}'$ and pressure P'. After the terms of second order in B' are dropped from equation (3.3.4), terms $\sim \xi B'$ and $\sim \xi P'$ appear; they can be interpreted as a random force and denoted by a random operator $\hat{\boldsymbol{K}}_c \boldsymbol{\xi}$. Equation (3.3.5) then becomes

$$\rho_0 \frac{\partial^2 \boldsymbol{\xi}}{\partial t^2} = -\hat{\boldsymbol{K}} \boldsymbol{\xi} - \hat{\boldsymbol{K}}_c \boldsymbol{\xi} \tag{3.3.6}$$

where

$$\begin{aligned} \hat{K}_c \boldsymbol{\xi} = & -(\gamma P' \operatorname{div} \boldsymbol{\xi} + \boldsymbol{\xi} \nabla P') \\ & - \frac{1}{4\pi}[\operatorname{rot} \operatorname{rot} [\boldsymbol{\xi}, \boldsymbol{B}'], \boldsymbol{B}_0] - \frac{1}{4\pi}[\operatorname{rot} \operatorname{rot} [\boldsymbol{\xi}, \boldsymbol{B}_0], \boldsymbol{B}'] \\ & - \frac{1}{4\pi}[\operatorname{rot} \boldsymbol{B}', \operatorname{rot} [\boldsymbol{\xi}, \boldsymbol{B}_0]] - \frac{1}{4\pi}[\operatorname{rot} \boldsymbol{B}_0, \operatorname{rot} [\boldsymbol{\xi}, \boldsymbol{B}']] \end{aligned}$$

and $(-\hat{\boldsymbol{K}} \boldsymbol{\xi})$ is equal, as before, to the right-hand side of equation (3.3.4).

To simplify the calculations, we set, without restricting generality, $P' = 0$, in order to avoid dealing with structurally identical (with respect to noise) additive terms in the expression for $\hat{\boldsymbol{K}}_c \boldsymbol{\xi}$; it will be clear soon that these additive terms do not affect the qualitative form of the solution. However, we do take them into account ($P' \neq 0$) in the cases in which density fluctuations play a decisive role (as in the first example in which perturbations of the interchange mode are studied); this corresponds to an effective random additional term $\rho_0 g'$ to the gravity force $\rho_0 g$.

In a similar manner, we can consider a random additional term to the magnetic field only ($\boldsymbol{B}' \neq 0$). To simplify the manipulations, we assume $\boldsymbol{B}'(\boldsymbol{r}, t)$ to be a random steady-state Gaussian quantity,

$$\langle \boldsymbol{B}' \rangle = 0 \qquad \langle \boldsymbol{B}'(\boldsymbol{r}, t) \boldsymbol{B}'(\boldsymbol{r}', t') \rangle = 2D(\boldsymbol{r}, \boldsymbol{r}', t) \delta(t - t')$$

where angle brackets denote the average value of a random variable over the ensemble of variable realization $\boldsymbol{B}'(\boldsymbol{r}, t)$, and $\delta(t)$ is the Dirac delta function. Choosing this time dependence of the random process $\boldsymbol{B}'(\boldsymbol{r}, t)$, we therefore assume in this problem that on the one hand, the characteristic variation times $t_B \gg \tau$ (τ is the correlation time of $\boldsymbol{B}'$) and on the other hand, the spectrum of the random variable $\boldsymbol{B}'(\boldsymbol{r}, t)$ contains resonant harmonics ω/n ($n = 1, 2, \ldots$) which make parametric destabilization possible. We know that a (3.3.6)-type equation describes small oscillations of an elastic medium with a generalized elasticity coefficient $\hat{\boldsymbol{K}}/\rho_0$ and a random additive term in frequency $\boldsymbol{K}_c/(\rho_0 K)$; it will be shown that this term is responsible for the parametric destabilization of the system. Taking into account that the operator $\boldsymbol{K}$ is self-conjugate, we can rewrite equation (3.3.6) as

$$u_i = \frac{\partial \xi_i}{\partial t} \qquad \rho_0 \frac{\partial u_i}{\partial t} = -\lambda_i \xi_i - K_{ijk}(B'_j \xi_k) \tag{3.3.7}$$

where λ_i are the eigenvalues of the operator $\hat{\boldsymbol{K}}$ and K_{ijk} is the tensor representation of the random operator $\hat{\boldsymbol{K}}$.

Averaging the set of equations (3.3.7) over the ensemble of realization of the random variable $\boldsymbol{B}'$, we obtain the following expressions for the mean displacement and mean velocity:

$$\langle u_i \rangle = \frac{\partial \langle \xi_i \rangle}{\partial t} \qquad \rho_0 \frac{\partial \langle u_i \rangle}{\partial t} = \lambda_i \langle \xi_i \rangle - K_{ijk} \langle B'_j \xi_k \rangle. \tag{3.3.8}$$

To close the set (3.3.8), we make use of the functional method (Klyatskin 1975). To do this, we find from (3.3.8) the expressions for the functional derivatives

$$\frac{\delta \xi_i(\boldsymbol{r}, t)}{\delta B'_m(\boldsymbol{r}', t')} = 0 \qquad \frac{\delta u_i(\boldsymbol{r}, t)}{\delta B'_m(\boldsymbol{r}', t')} = -K_{ijk} \delta_{mj} \xi_k(\boldsymbol{r}, t) \delta(\boldsymbol{r} - \boldsymbol{r}') \delta(t - t')$$

where δ_{mj} is the Kronecker symbol.

After the application of the Furutsu–Novikov formula (Klyatskin 1975), the set (3.3.3) becomes closed because the last term in the second equation equals zero; note that for first moments the set (3.3.8) is indistinguishable from the conventional system of MHD equations with fluctuations neglected. Therefore, let us consider one-point second moments of system (3.3.7):

$$\begin{gathered} \langle u_i u_j \rangle = \left\langle \frac{\partial \xi_i}{\partial t} u_j \right\rangle \qquad \langle u_i \xi_j \rangle = \left\langle \frac{\partial \xi_i}{\partial t} \xi_j \right\rangle \\ \left\langle \frac{\partial u_i}{\partial t} u_j \right\rangle = -\lambda_i \langle \xi_i u_j \rangle - K_{imk} \langle B'_m \xi_k u_j \rangle \\ \left\langle \frac{\partial u_i}{\partial t} \xi_j \right\rangle = -\lambda_i \langle \xi_i \xi_j \rangle - K_{imk} \langle B'_m \xi_k \xi_j \rangle. \end{gathered} \tag{3.3.9}$$

Using the Furutsu–Novikov formula, we find for the correlators

$$K_{imk} \langle B'_m \xi_k u_j \rangle = -K_{imk} D_{mp} K_{jpq} \langle \xi_k \xi_q \rangle \qquad K_{imk} \langle B'_m \xi_k \xi_i \rangle = 0.$$

The closed set of equations obtained above is too complex for analysis. We can use a simplifying assumption. In many plasma stability problems one considers symmetric configurations, so that the problem reduces to formally one-dimensional dependences; in our case this is equivalent to diagonal correlation matrices. Under this assumption, the equations for one-point second moments (3.3.9) reduce to the form

$$\begin{aligned}
&\frac{\partial}{\partial t}\langle \xi u\rangle = \langle u^2\rangle - \frac{1}{\rho_0}\langle \xi^2\rangle \\
&\frac{1}{2}\frac{\partial}{\partial t}\langle u^2\rangle = -\frac{\lambda}{\rho_0}\langle \xi u\rangle + \frac{(K_c)^2}{\rho_0}D\langle \xi^2\rangle \\
&\frac{1}{2}\frac{\partial}{\partial t}\langle \xi^2\rangle = \langle \xi u\rangle .
\end{aligned} \tag{3.3.10}$$

Equation (3.3.10) yields the following dispersion equation for perturbations $\sim \exp(\mathrm{i}\omega t)$:

$$\omega^3 - \frac{\lambda}{\rho_0}\omega - \mathrm{i}\frac{(K_c)^2}{\rho_0}D = 0. \tag{3.3.11}$$

If we assume $D = 0$ in (3.3.11), we have an ordinary dispersion equation of plasma oscillations whose solution is $\omega \approx -\mathrm{i}(K_c)^2 D/(2\lambda)$, or an unstable equilibrium state (with $\lambda < 0$). If small fluctuations are present ($D \neq 0$, $D \ll 1$), equation (3.3.11) has a new solution $\omega_{1,2}^2 = \lambda/\rho_0$, which points to the onset of instability. If, however, the state is unstable, it is readily seen that the growth rate grows with increasing noise.

Taking pairwise collisions into account results, in the simplest case, in a new term $-\rho_0\nu(\partial\xi/\partial t)$ in the right-hand side of (3.3.4) (ν being the collision frequency). In this case the instability will arise only when the noise signal power satisfies the condition

$$4K_c^2 D > \nu\rho_0(4\lambda - \nu^2\rho_0)$$

and its growth rate is $[Dk_c^2 - \nu\rho_0^2(\lambda/\rho_0 - \nu^2/4)]/[\rho_0^2(\lambda/\rho_0 - \nu^2/4)]$.

It will be shown below that intermittency may exist in a system of the type of (3.3.7). To achieve this, we rewrite (3.3.7) in view of the remarks made above:

$$u = \dot{\xi} \qquad \dot{u} = -\lambda^2(1 + q(t))\xi \tag{3.3.12}$$

where $q(t)$ is the random Gaussian quantity, $\langle q\rangle = 0$ and $\langle qq'\rangle = 2D\delta(t - t')$. Equation (3.3.12) yields a set of equations for moments of order n ($n = 1, 2, \ldots$):

$$\begin{aligned}
\frac{\partial}{\partial t}\langle u^p \xi^{(n-p)}\rangle = &-p\lambda^2\langle u^{(p-1)}\xi^{(n-p+1)}\rangle + p(p-1)D\lambda^4\langle u^{(p-2)}\xi^{(n-p+2)}\rangle \\
&+ (n-p)\langle u^{(p+1)}\xi^{(n-p+1)}\rangle \qquad p = 0, 1, \ldots, n
\end{aligned}$$

As moments of different order are seen to be coupled in the resulting set of equations, we will consider the following representation for variables:

$$\begin{aligned} \xi(t) &= A(t)\sin(\lambda t + \varphi(t)) \equiv A(t)\sin(\Phi(t)) \\ u(t) &= \lambda A(t)\cos(\lambda t + \varphi(t)) \equiv \lambda A(t)\cos(\Phi(t)). \end{aligned} \tag{3.3.13}$$

Then (3.3.12) gives us Φ and v after the substitution $A = \mathrm{e}^{v}$:

$$\begin{aligned} \dot{\varphi} &= \lambda q \sin^2 \Phi \\ \dot{v} &= -\frac{\lambda}{2} q \sin 2\Phi. \end{aligned} \tag{3.3.14}$$

The Einstein–Fokker equation corresponding to (3.3.13) yields as a solution, after averaging over a period $2\pi/\lambda$ (we assume here $\lambda > 0$), the distribution $P_t(v)$ that was obtained by Kolomiets (1962)

$$P_t(v) = \frac{1}{\sqrt{2\pi}\sigma(t)} \exp\left(-\frac{1}{2}\left(\frac{v - \langle v \rangle}{\sigma(t)}\right)^2\right)$$

where

$$\langle v \rangle = v_0 + \frac{D\lambda^2 t}{4} \qquad \text{and} \qquad \sigma(t) = \frac{D^2\lambda^2}{4} t.$$

The central moment of order n for the exponent in the expression for oscillation amplitude is given by

$$\langle (v - \langle v \rangle)^n \rangle_t = \int (v - \langle v \rangle)^m P_t(v)\,\mathrm{d}v.$$

Since the distribution of v is Gaussian, the moment is zero for odd n. For even n we find

$$\langle (v - \langle v \rangle)^n \rangle_t = (1 \cdot 3 \cdot 5 \ldots (n-1))\sigma^n \equiv (n-1)!!\sigma^n.$$

Equation (3.3.4) shows that the rate of growth of the nth moment is of the order of $(n-1)nD^2\lambda^2/4$ and increases steeply with increasing order of the moment. As shown by Molchanov *et al* (1985), the increase in the growth rate with increasing moment order indicates that intermittency is possible in the dynamic system. In other words, the increasing-amplitude field may be shaped into surges of field, contracting and growing in amplitude. Of course, this unusual 'raking up' behavior also characterizes, of course, the magnetic field in the case of non-potential instabilities. Note that we have discussed the behavior of a linear system and thus dynamic chaos cannot develop in principle, while chaos due to fluctuations grows exponentially because dispersion grows exponentially (on dynamic chaos, see, for example, Schuster 1984). This conclusion is of principal importance. First, small fluctuations are responsible for the exponential growth rate in the stability range of completely integrable systems. Second, even a small addition to the growth rate in the range of instability results in irregular behavior of the system even in the linear approximation; hence at least the transition to stochastic instability, which starts the dynamic chaos, must be analyzed against the background of the evolving fluctuational chaos. We will deal with this aspect later; here we look at the linear stage in two specific examples.

A. Interchange (flute) perturbations of the plasma. Let us consider in detail a specific example of the effect of fluctuations on interchange perturbations in plasmas. As we know, the equation describing the interchange modes of the electric potential ψ in the case where there is a non-perturbed gradient of the density n_0 in the direction perpendicular to the external magnetic field $H_0 \equiv H_{0z}$, has the form

$$\ddot{\Psi} + gk_0\Psi = 0. \tag{3.3.15}$$

Here g is the acceleration of the effective gravitational field that imitates the effect of curvature of the field lines $g = v_T^2/R_c$, where v_T is the thermal velocity of particles, R_c the curvature radius of a field line, $k_0 = -(1/n_0)\,\mathrm{d}n_0/\mathrm{d}x$, and the axis OX is directed along the inhomogeneity of n.

Let us assume that against the background of the average steady-state effective gravitational field $g_c(\boldsymbol{r}, t)$, there is a fluctuational part $\tilde{g}(\boldsymbol{r}, t)$, relatively small in comparison, caused by the random oscillation of pressure (for example, due to small turbulence fluctuations (pulsations)), that is,

$$g(\boldsymbol{r}, t) = g_c(\boldsymbol{r}, t) + \tilde{g}(\boldsymbol{r}, t). \tag{3.3.16}$$

If $\tilde{g} = 0$, then (3.3.15) describes instability (at $gk_0 < 0$) with growth rate $\gamma = \sqrt{|gk_0|}$.

Taking (3.3.16) into account, we can rewrite equation (3.3.15) as a set of equations

$$\theta = \dot{\Psi} \qquad \dot{\theta} = -g_c k_0 \Psi - \tilde{g} k_0 \Psi.$$

After closing and averaging, by analogy to the scheme described previously, we obtain for second moments the dispersion equation

$$\omega^3 - 4g_c k_0 \omega - 4\mathrm{i}D(\boldsymbol{r}, \boldsymbol{r})k_0 = 0. \tag{3.3.17}$$

Traditionally, $\tilde{g}$ is a random Gaussian quantity and $\langle \tilde{g} \rangle = 0$, so that

$$\langle \tilde{g}(\boldsymbol{r}, t)\tilde{g}(\boldsymbol{r}', t') \rangle = D(\boldsymbol{r}, \boldsymbol{r}', t)\delta(t - t').$$

In the case when $g_c k_0 > 0$ and the instability cannot set in without fluctuations, it is easily shown that the parametric destabilization results in an instability with the growth rate given by

$$\gamma = Dk_0/g_c.$$

In the other case, when $g_c k_0 < 0$, we set $\omega = \omega_0 + \tilde{\omega}$, where ω_0 is the solution of the dispersion equation for the interchange mode in the absence of noise, and $\tilde{\omega}$ is a small quantity added to the frequency, which is a consequence of the fluctuational term in equation (3.3.17). For $\tilde{\omega}$ we then have the expression $\tilde{\omega} = 4\mathrm{i}Dk_0/g_c k_0$. Obviously, this addition to ω_0 at $g_c k_0 < 0$ accelerates the onset of instability. It is not difficult to deduce that the accelerated mode of instability appears here because of the dominant effect of the random 'convexities' of the magnetic field.

If ω is sufficiently high and satisfies the inequality $\omega\tau_0 \ll 1$, where τ_0 is the characteristic correlation radius of the random variable $\tilde{\gamma}$, there is an unstable root $\gamma = \sqrt[3]{4Dk_0^2}$. In this case the fluctuational chaos develops so rapidly that speaking about the dynamic chaos becomes meaningless.

B. Finite-conductivity plasma. For a second example, we consider a model problem of fluctuational parametric instability of a plasma with finite conductivity. Let the plasma be placed in a uniform very high magnetic field $\boldsymbol{B}$. Let OZ be directed along $\boldsymbol{B}$ and let us assume that a weak current $\boldsymbol{j}$ flows in this direction. For simplicity, we assume that pressure in the plasma is negligible compared with that of the magnetic field, and also assume that plasma conductivity σ_0 in the equilibrium state is a slowly varying function of x. Small perturbations of electric field are assumed vortex-free, so that the linearized set of equations becomes (Kadomtsev 1963)

$$\begin{aligned} &\rho_0 \frac{\partial \boldsymbol{v}}{\partial t} = \frac{1}{c}[\boldsymbol{j}, \boldsymbol{B}] \qquad && \boldsymbol{E} = -\nabla\varphi \\ &\boldsymbol{j} = \sigma_0\left(\boldsymbol{E} + \frac{1}{c}[\boldsymbol{v}, \boldsymbol{B}]\right) && \frac{\partial \sigma}{\partial t} + v_x \frac{\partial \sigma_0}{\partial x} = \chi k_z^2 \sigma \end{aligned} \tag{3.3.18}$$

where ρ_0 is the plasma density, χ is the electric conductivity coefficient along the field lines, c is the velocity of light, and perturbations are proportional to $\exp(\mathrm{i}\boldsymbol{k}\boldsymbol{r})$.

It is not difficult to derive from (3.3.18) the equation for small velocity perturbations along the OX-axis (in what follows, we simply denote perturbations by v):

$$\frac{\partial^2 v}{\partial t^2} + \frac{\sigma_0 B^2}{\rho_0 c^2}\left(1 + \frac{\chi k_z^2 \rho_0 c^2}{\sigma_0 B^2}\right)\frac{\partial v}{\partial t} + \frac{\sigma_0 B^2}{\rho_0 c^2}\left(\chi^2 k_z - \frac{E_0 c k_y}{B k_z \sigma_0}\frac{\mathrm{d}\sigma_0}{\mathrm{d}x}\right) v = 0. \tag{3.3.19}$$

When random fluctuations of magnetic field are taken into account, a term appears in equation (3.3.19) acting as stochastic negative friction; however, we will not consider this case now since our goal here is to analyze parametric destabilization. Assume that the electric field E_0 in (3.3.19) includes random oscillations E' over the fixed $\bar{E}_0$, that is $E_0 = \bar{E}_0 + E'$ and $\langle E' \rangle = 0$; $\langle E'(x,t)E'(x',t')\rangle = 2D(x,x',t)\delta(t-t')$ is a steady-state Gaussian quantity.

Hence this will not affect first moments and we will arrive at the following dispersion relation for the perturbation $\sim \exp(\omega t)$ from the equations for second moments of the quantities v and $u \equiv \partial v/\partial t$:

$$(\omega + \eta)^3 + (4\omega_0^2 - \eta^2)(\omega + \eta) - 4a^2 D = 0 \qquad (3.3.20)$$

where

$$\eta = \frac{\sigma_0 B^2}{\rho_0 c^2} + \chi k_z^2$$

$$\omega_0^2 = \left(\chi k_z^2 - \frac{k_y E_0 c}{k_z B_0}\frac{\mathrm{d}\sigma_0}{\mathrm{d}x}\right)\frac{\sigma_0 B^2}{\rho_0 c^2}$$

$$a^2 = \frac{\sigma_0 B k_y}{\rho_0 c k_z}.$$

Equation (3.3.20) shows that at sufficiently high values of k_y there is an unstable root

$$\omega = 4\left(\frac{\sigma_0 B k_y}{\rho_0 c k_z}\right)^{2/3} D^{1/3} - \frac{\sigma_0 B^2}{\rho_0 c^2} - \chi k_z^2;$$

the instability condition is

$$4D\left(\frac{k_y}{k_z}\right)^2 > \frac{\sigma_0 B_0^2}{\rho_0 c^2}\left(1 + \frac{\chi k_z^2 c^2 \rho_0}{\sigma_0 B_0^2}\right)^3.$$

Considering the effect of finite conductivity leads to more stringent requirements to the evolution of the instability: unstable modes then set in at any level of dispersion of external noise above a certain non-zero threshold value.

This analysis allows us to formulate the following conclusions:

(i) if fluctuations are taken into account, MHD equilibrium of the plasma is universally unstable in the range where it is stable in the absence of fluctuations; instabilities arise due to the parametric destabilization of the natural plasma oscillations.
(ii) As a result of fluctuations, the growing plasma oscillations may become intermittent.
(iii) The fluctuational chaos setting in the system results in the stochastic behavior of plasma and an energy transfer as early as the linear stage.
(iv) In the case of instability, multiplicative noise accelerates the transition to instability.
(v) The noise that is external relative to a given subsystem may result in the appearance of terms that give a negative contribution to dissipative loss (for sign reversal in the friction coefficient, see also Leibowitz 1963).

The difference between the effects of regular parametric perturbation and of stochastic perturbation are essential. If parameters vary in a regular manner, instabilities arise only in specific regions of space while random fluctuations of parameters cause instability no matter how small the fluctuation level.

The aspects discussed above are of practical importance for plasma physics; however, there are also various principal problems whose analysis must change dramatically if the growing fluctuational perturbation is taken into account. A special place among such problems is occupied by self-organization, that is, the formation of structures; this process starts, as we well know, with instability onset.

In view of these arguments, we will analyze in this section the effect of multiplicative fluctuations on systems close to their stationarity points. Using then the subordination principle (see, for example, Haken 1978), we obtain the following equation for the ordering parameter y:

$$\frac{\mathrm{d}y}{\mathrm{d}t} = (\Lambda_1 y + Ay^3) + (z_1 y + z_2 y^2). \tag{3.3.21}$$

Let a system be stable in the absence of fluctuations. The equation for the ordering parameter Λ then includes the eigenvalue of the matrix that determines the stability of the linear problem, that is, $(\mathrm{Re}\Lambda_1) < (\mathrm{Re}\Lambda_i)$ $(i \neq 1)$, where i characterizes the sequence of equations of the type (3.3.21) that describe the state of the system close to a singular point. In (3.3.21) we have also taken into account the cubic nonlinearity and the characteristic functions z_1 and z_2 that are statistically independent ($\langle z_1 z_2 \rangle = 0$). We also assume that the ordinary relations $\langle z_j \rangle = 0$, $\langle z_j(t) z_j(t') \rangle = 2D_j \delta(t - t')$, $j = 1, 2$ (no summation over j) hold. Using the Furutsu–Novikov formula (see, for example, Klyatskin 1975), we obtain the following set of equations for the moments:

$$\frac{\mathrm{d}}{\mathrm{d}t}\langle y^n \rangle = n(\Lambda_1 + nD_1)\langle y^n \rangle + n[A + (n+1)D_2]\langle y^{n+2} \rangle. \tag{3.3.22}$$

Equation (3.3.22) implies a number of important results, including the conclusion that taking fluctuations into account destabilizes the moments. The presence of D_2 may trigger an explosive growth of moments. First we consider some consequences in the linear approximation ($A = D_2 = 0$). Then (3.3.22) shows that fluctuations invariably weaken stability if $\mathrm{Re}\Lambda_1 < 0$, and accelerate the evolution of instability if $\mathrm{Re}\Lambda_1 > 0$. Furthermore, even though random forces are small, there is always such a number n that $nD_1 > |\mathrm{Re}\Lambda_1|$, that is, in the vicinity of singular points random forces destabilize moments if $n > |\mathrm{Re}\Lambda|/D_1$; note that higher-order moments grow at higher growth rates. Molchanov *et al* (1985) showed that this peculiarity in moment growth gives ground for suggesting intermittency (see the references cited therein for a bibliography of earlier publications on intermittency). We can thus formulate the following general result: the presence of multiplicative random forces in the vicinity of a bifurcation point always produces accelerated evolution of the system and may result in intermittent stochasticity. Note that a similar result was discovered by analyzing near-threshold modes of stratified shear flows (Moiseev *et al* 1982, 1984). As an example of using (3.3.22) in a nonlinear mode, consider the dynamics of the second moment. Assuming deviation y from the bifurcation point to be small and using the Gaussian approximation, we express the fourth moment $\langle y^4 \rangle$ in terms of the second one. Introducing the notation $\Lambda^* \equiv (\Lambda_1 + 2D_1)$ and $A^* = (A + 3D_2)$ and assuming $\Lambda^* < 0$ and $A^* > 0$, we derive from (3.3.22)

$$\langle y^2 \rangle = \frac{\langle y^2 \rangle_0 \exp(-2t|\Lambda^*|)}{1 - (A^* \langle y^2 \rangle_0 / |\Lambda^*|)\left[1 - e^{-2t|\Lambda^*|}\right]} \tag{3.3.23}$$

where $\langle y^2 \rangle_0$ is the initial value of the second moment.

Formula (3.3.23) implies that if inequality $A^* \langle y^2 \rangle_0 > |\Lambda^*|$ is satisfied, the second moment grows with time 'explosively', even though the linear approximation implies damping of this growth.

The results outlined in this section illustrate the strong effect of fluctuations on the stability of dynamic systems (and hence, the formation of structures, the heating, diffusion and acceleration of particles), including plasma-like systems. In the next section we show, for a specific situation, that their role increases still further if we take into account the resonance processes in the plasma.

3.4 Anomalous diffusion of particles

Following Butz and Moiseev (1990), we consider the dynamics of a charged particle in an external magnetic field whose constant component H_0 points along the z-axis and a fluctuating component $\tilde{H}(t)$ also directed along this axis. Assume that in addition to the magnetic field, there is also the field of an external plane electromagnetic wave that propagates along the z-axis. In this case the behavior of a charged particle is characterized by the familiar integral of motion (see, for example, Davydovskii 1962 and Kolomyanskii and Lebedev 1962)

$$I \equiv \frac{v_z}{c}\gamma = \kappa\gamma + \text{const}.$$

Here v_z and γ are, respectively, the projection of the particle's velocity on z and its relativistic factor; $\kappa = (kc)/\omega$ (ω, k are the frequency and wave vector of the wave). We are interested now in the case of efficient interaction of a particle with the field close to one of the resonance conditions in constant field H_0,

$$R_{n0} \equiv \frac{kv_{z0}}{\omega} + \frac{n\omega_{H_0}}{\gamma\omega} - 1 = 0 \tag{3.4.1}$$

(R_{n0} vanishes at exact resonance). In (3.4.1) n is the resonance number, ω_{H_0} is the cyclotron frequency in the field (here and further on we assume $\tilde{H}$ to be a small corrective term). We emphasize that if $\kappa \to 1$ and $I \to -n(\omega_{H_0}/\omega)$, the integral of motion coincides with the condition of cyclotron resonance (the case of automatic resonance, or autoresonance). Overlapping of resonances produces dynamic chaos, particles begin to diffuse in the space of energy, and unlimited acceleration of charged particles becomes possible. However, the acceleration rate is not too high and the phase space always contains stability islands; if a particle reaches an island, it stays in it for a long time. Magnetic field fluctuations $\tilde{\omega}_H$ cause particle diffusion, which dominates behavior in stability islands and in isolated resonances when the dynamic chaos is not produced. In what follows we will limit the analysis to a single isolated resonance and assume that the change $\tilde{\gamma}$ in particle energy caused by the wave and the fluctuating field is small ($\tilde{\gamma} \ll \gamma_0$, where γ_0 is the initial relativistic factor). Now we choose an approximation linear in $\tilde{\gamma}$, use the equations of motion of a particle in this particular case, and obtain

$$\begin{aligned} \frac{\mathrm{d}\tilde{\gamma}_n}{\mathrm{d}\tau} &= \frac{1}{\gamma_0} E_0 W_n \cos\theta_n \\ \frac{\mathrm{d}\theta_n}{\mathrm{d}\tau} &= \left.\frac{\partial R_{no}}{\partial \gamma}\right|_{\gamma=\gamma_0} \tilde{\gamma}_n + \frac{n\tilde{\omega}_H}{\gamma_0} \\ \theta_n &\equiv kz - n\theta - \tau \end{aligned} \tag{3.4.2}$$

where θ is the angle between the particle's momentum and the x-axis, $E_0 = (eE_0)/(mc\omega)$ (e is the particle charge, E_0 is the amplitude of the electric vector of the field strength in the wave), $\tau = \omega t$,

$$W_n \equiv \alpha_x p_\perp \frac{n}{\mu} J_n(\mu) - \alpha_y p_\perp J_n'(\mu)$$

α_x and α_y are the projections of the polarization vector on the axes x and y, respectively, $p_\perp$ is the projection of the particle's momentum on the z-axis, $\mu \equiv p_\perp/\omega_{H_0}$ and J_n is the Bessel function.

Note that if $x \to 1$ and $v_z \to c$, we can neglect the changes in $p_\perp$. Equations (3.4.2) describe the motion of a mathematical pendulum whose phase is affected by an external fluctuational force $\sim \tilde{\omega}_H$. Omitting for the moment the fluctuational term, we study the structure of the nonlinear resonance. Equations (3.4.2) imply the following width of an isolated nonlinear resonance:

$$\Delta\frac{\mathrm{d}\theta_n}{\mathrm{d}\tau} = 4\sqrt{\left.\frac{\partial R_{no}}{\partial\gamma}\right|_{\gamma_0} \frac{E_0 W_n}{\gamma_0}}. \tag{3.4.3}$$

We are mostly interested in the resonance structure in the energy space. Using (3.4.3), we find from (3.4.2) the following nonlinear resonance width in energy units:

$$\Delta\tilde{\gamma}_n = 4\sqrt{E_0 W_n/\gamma_0 \left.\frac{\partial R_{no}}{\partial\gamma}\right|_{\gamma_0}}. \tag{3.4.4}$$

Let us point to one important feature of equation (3.4.4): as the derivative of resonance with respect to energy $\partial R_{no}/\partial\gamma|_{\gamma_0}$ decreases, the nonlinear resonance width increases. The tendency of the derivative to zero signifies that the resonance is sustained as a particle interacts with the wave, that is, the resonance relation is one of the integrals of motion. It is not difficult to see that as $\partial R/\partial\gamma \to 0$, the autoresonance is produced (see above). It can also be shown that as individual resonances get wider in nearly-autoresonance conditions, the neighboring resonances 'recede' from one another, that is, we look at a case that is an opposite of dynamic chaos, and so it is especially important to take fluctuations into account.

Let us look now at the peculiarities of motion of a charged particle affected by fluctuations of the external magnetic field. To achieve this, it will be sufficient to consider the linearized set of equations (3.4.2),

$$\begin{aligned}\frac{\mathrm{d}\gamma}{\mathrm{d}\tau} &= -B\theta \\ \frac{\mathrm{d}\theta}{\mathrm{d}\tau} &= \alpha y + f\end{aligned} \tag{3.4.5}$$

where $B \equiv E_0 W_n/\gamma_0$, $\theta = \theta_n - n/2$, $\alpha \equiv \partial R/\partial\gamma\big|_{\gamma_0}$ and $f \equiv n\tilde{\omega}_n/\gamma_0$, $y = \tilde{\gamma}$.

We assume now that $\langle f\rangle = 0$, $\langle f(\tau)f(\tau')\rangle = 2D\delta(\tau - \tau')$. Then, by analogy to the previous sections, we easily obtain

$$\begin{aligned}\frac{\mathrm{d}}{\mathrm{d}\tau}\langle\gamma\rangle &= -B\langle\theta\rangle \\ \frac{\mathrm{d}}{\mathrm{d}\tau}\langle\theta\rangle &= \alpha\langle\gamma\rangle \\ \frac{\mathrm{d}}{\mathrm{d}\tau}\langle\gamma^2\rangle &= -2B\langle\gamma\theta\rangle \\ \frac{\mathrm{d}}{\mathrm{d}\tau}\langle\theta\gamma\rangle &= -B\langle\theta^2\rangle + \alpha\langle\gamma^2\rangle \\ \frac{\mathrm{d}}{\mathrm{d}\tau}\langle\theta^2\rangle &= 2\alpha\langle\gamma\theta\rangle + 2D.\end{aligned}$$

It follows from these equations that

$$\langle\gamma^2\rangle = \frac{E_0 W_n D}{2\gamma_0(\partial R/\partial\gamma)}\tau.$$

Therefore, even though the resonance width grows anomalously rapidly as $(\mathrm{d}R)/(\mathrm{d}\gamma) \to 0$, the diffusion of a charged particle inside the resonance is enhanced even faster. The effect of fluctuations is felt even stronger in resonance diffusion, that is, if a particle diffuses together with the resonance. Set (3.4.2) describes the 'wandering' resonance centered at $\tilde{\gamma} = -(n\omega_H/\gamma_0)/(\partial R/\partial\gamma)$. This gives us the following expression for the local diffusion coefficient:

$$\langle\tilde{\gamma}^2\rangle = 2D_\lambda\tau \qquad D_\lambda = \frac{n^2 D}{\gamma_0^2(\partial R/\partial\gamma)^2}.$$

The fluctuational diffusion under autoresonance can thus greatly exceed the ordinary quasi-linear diffusion.

Note that the role of fluctuations in the wave parameters (amplitude or phase) can also be found from set (3.4.5). Among other things, we find from it the following expression for the second momentum:

$$\langle \gamma^2 \rangle \sim \exp \left[\sqrt[3]{2B^2 D} \left(\frac{\partial R_n}{\partial \gamma} \bigg|_{\gamma_0} \right)^{2/3} \tau \right]. \tag{3.4.6}$$

Equation (3.4.6) shows that the effect of wave fluctuation parameters becomes predominant far from the autoresonance conditions.

3.5 Fluctuations of transfer coefficients in the plasma and in hydrodynamics

In this section we follow Moiseev *et al* (1990) and concentrate specifically on the fluctuations of those parameters of a system that enter the equations of its dynamics, that is, the coefficient with the derivative with respect to time of order one less than that of the leading derivative. Note that if such fluctuations are taken into account, the system's dynamics is described by stochastic differential equations (SDE) that can be used as basic mathematical models for a broad class of problems in physics, biology, economics and chemistry. The simplest illustration is the one-dimensional motion of a particle of unit mass subjected to a fluctuating friction coefficient $\gamma = \gamma_0 + \tilde{\gamma}(t)$, $\langle \tilde{\gamma} \rangle = 0$, described by the stochastic differential equation

$$\ddot{r} + (\gamma_0 + \tilde{\gamma})\dot{r} + V_r'(r) = 0 \tag{3.5.1}$$

in which the derivatives are treated in the generalized interpretation and $V(r)$ is the particle potential. After averaging over the realization ensemble $\tilde{\gamma}(t)$ with the correlation function $\langle \tilde{\gamma}(t_1)\tilde{\gamma}(t_2) \rangle = D(t_2 - t_1)$, equation (3.5.1) can be rewritten in the form

$$\langle \ddot{r} \rangle + (\gamma_0 - \gamma(D))\langle \dot{r} \rangle + \langle V_r'(r) \rangle = 0 \tag{3.5.2}$$

where $\gamma(D) > 0$ and $\gamma(D) \sim D$; if the correlation time of the Gaussian quantity $\tilde{\gamma}(t)$ tends to zero, we can always use the approximation $D(t_2 - t_1) = D\delta(t_2 - t_1)$ and then obtain from (3.5.2)

$$\langle \ddot{r} \rangle + (\gamma_0 - D)\langle \dot{r} \rangle + \langle V_r' \rangle = 0. \tag{3.5.3}$$

Clearly, averaged equations (3.5.2) and (3.5.3) describe a system with the value of the effective friction coefficient reduced as a result of interaction with noise; 'negative friction' arises if $D > \gamma_0$. Mikhailov and Uporov (1984) showed that the condition $D = \gamma_0$ in a similar equation describing the reproduction of some matter determines the threshold above which explosive instability sets in.

The following argument clarifies this unexpected result. The realizations $\tilde{\gamma}(t)$ ($\tilde{\gamma} > 0$) for which the damping rate of the particle velocity $V = \dot{r}$ becomes so high that V rapidly drops to zero do not contribute to the average value $\langle V \rangle$. If the distribution function $\tilde{\gamma}(t)$ is symmetric relative to zero, there is the same probability for the values $\tilde{\gamma} < 0$ of the same magnitude but at which the damping rate of velocity and V will be much slower. If damping starts from the same level as in the preceding case, such realizations yield the main contribution to the average value $\langle V \rangle$. This logic also holds for non-Gaussian distributions $\tilde{\gamma}$ that are symmetric with respect to zero. The effect of appearance of stochastic 'negative friction' was mentioned by Leibowitz (1963).

Note that the non-equilibrium noise-induced phase transition in the Verhulst model that describes the biological population growth (Horsthemke and Malek-Mansour 1976) is similar to the sign reversal of friction in (3.5.3).

Finding the effect of fluctuations of the viscosity coefficient on a system's evolution belongs to the same type of problem. Viscosity fluctuations occur in multicomponent systems with reactions, and also in the case of a strong temperature dependence of kinematic viscosity and in systems within their critical regions (Mikhailov and Uporov 1984, Betchov and Criminale 1967). Fluctuations of turbulent viscosity (we know that turbulent viscosity depends on the energy of turbulent motion) simulate the effect of intermittency on transfer processes. Note that Kraichnan (1976) analyzed a similar problem with helicity fluctuations in magnetic hydrodynamics.

To simplify the manipulations, consider the equation of diffusion of a scalar admixture (the conclusions are unchanged for a vector admixture),

$$\frac{\partial C}{\partial t} = \chi \Delta C \tag{3.5.4}$$

where the diffusion coefficient is $\chi = \chi_0 + \tilde{\chi}(t)$, $\langle \tilde{\chi}(t)\tilde{\chi}(t') \rangle = 2D_f \delta(t - t')$.

We ignore here the specific nature of viscosity fluctuations. To simplify further, we consider only δ-correlated Gaussian time fluctuations.

Note that similar results are obtained for short-range-correlated and spatially distributed fluctuations.

Having averaged (3.5.4) over short-duration fluctuations, we readily arrive at the equation

$$\frac{\partial}{\partial t}\langle C \rangle = \chi_0 \Delta \langle C \rangle + D_f \Delta^2 \langle C \rangle. \tag{3.5.5}$$

It is clear from (3.5.5) that this equation differs from an ordinary diffusion equation in an additional term on the right-hand side, which conserves the amount of both the scalar and the vector admixture but greatly changes its spatial distribution as $t \to \infty$. While the ordinary diffusion equation spreads the admixture with time over the entire space, equation (3.5.5) describes its small-scale localization. It can be expected that admixture distribution will become intermittent as a result of diffusion with fluctuating coefficient, that is, the arbitrary initial distribution of admixture evolves to a state in which the admixture is concentrated at isolated, randomly distributed centers. If fluctuations of the diffusion coefficient are related to turbulence intermittency, then non-Gaussian characteristics of turbulence can be evaluated from admixture distribution at $t \to \infty$.

References

Barts B I, Lapidus I I and Moiseev S S 1988 *Preprint No 1375* (Moscow: Space Research Institute) pp 1–42 (in Russian)

Bernstein I B, Frieman E A, Kruskal M D and Kulsrud R M 1958 *Proc. R. Soc.* **A244** 17

Betchov R and Criminale W O Jr 1967 *Stability of Parallel Flows* (New York: Academic)

Butz V A and Moiseev S S 1990 *Zh. Tech. Fiz.* **60** 35–42

Butz V A, Moiseev S S and Shavva I I 1989 *Contr. Papers to 1989 Int. Conf. on Plasma Phys. (New Delhi, 1989)* vol 2, ed A Sen and P K Kaw pp 349–51

Butz V A, Moiseev S S and Shavva I I 1990 *Fizika Plazmy* **16** 771–8

Davydovskii V Ya 1962 *Zh. Exp. Teor. Fiz.* **43** 886–8

Gaponov-Grekhov A V and Rabinovich M I (eds) 1983 *Nelineynye Volny: Samoorganizatsiya* (Moscow: Nauka) pp 1–263 (in Russian)

Haken H 1978 *Synergetics* (Berlin: Springer)

Horsthemke W and Lefever R 1984 *Noise-Induced Transitions* (Berlin: Springer)

Horsthemke M and Malek-Mansour M 1976 *Zs. Phys.* **B24** 307

Kadomtsev B B 1963 *Voprosy Teorii Plazmy Vyp. 2* (Moscow: Gosatomizdat) p 132, in Russian

Klyatskin V I 1975 *Statisticheskoe Opisanie Dinamicheskih System s Fluktuiruyuschimi Parametrami* (Moscow: Nauka) pp 1-239 (in Russian)

Kolomiets V G 1962 *Ukr. Mat. Zh.* **14** 211

Kolomyanskii A A and Lebedev A K 1962 *Dokl. Akad. Nauk SSSR* **145** 1259–61

Kraichnan R H 1976 *J. Fluid Mech.* **77** 753–68

Leibowitz M A 1963 *J. Math. Phys.* **4** 852–8

Mikhailov A S and Uporov I V 1984 *Usp. Fiz. Nauk* **144** 79–112

Moiseev S S, Pungin V G, Sagdeev R Z, Suyazov N V and Etkin V S 1982 *Abstract 2nd Congress Soviet Oceanologists (Sevastopol, 1982)* pp 63–4

Moiseev S S, Suyazov N V and Etkin V S 1984 *Preprint No 905* (Moscow: Space Research Institute) pp 1–19 (in Russian)

Moiseev S S, Chkhetiani O G and Shavva I I 1990 *Preprint No 1667* (Moscow: Space Research Institute) pp 1–15 (in Russian)

Molchanov S A, Ruzmaikin A A and Sokolov D D 1985 *Sov. Phys.–Usp.* **28** 307

Neimark Yu I and Landa P S 1987 *Stohasticheskie i Haoticheskie Kolebaniya* (Moscow: Nauka) pp 1–424 (in Russian)

Schuster H G 1984 *Deterministic Chaos* (Weiheim: Physik-Verlag)

Shafranov V D 1963 *Voprosy Teorii Plazmy* vol 3, ed M A Leontovich (Moscow: Gosatomizdat) pp 3–140 (in Russian)

Zaslavskii G M 1984 *Stokhastichnost Dinamicheskih System* (Moscow: Nauka) pp 1–271 (in Russian)

Zaslavskii G M and Moiseev S S 1966 *Zh. Tech. Fiz.* **36** 2217

Chapter 4

Non-standard evolution of turbulent media

4.1 Introduction

In this chapter we outline another approach to developing the concept of secondary instabilities. The level of perturbations generated as a result of primary instability or external noise was considered invariant in the course of evolution of the secondary instability. However, as secondary perturbations grow in amplitude, their feedback effect on the primary perturbations grows as well, so that further stages of this instability must be studied in a self-consistent manner. This problem is much more complex. However, as we show in more detail in chapter 5, it is extremely important for understanding a number of geophysical processes in various shells of the Earth, including its crust, hydrosphere, atmosphere and ionosphere. Hence even model examples, helping to improve our understanding of these interactions, are of considerable interest.

Section 2 of this chapter discusses, as an example, one of the possible ways of evolution of turbulence of drift waves in spatially non-uniform weakly ionized plasma placed in crossed external electric and magnetic fields, so that the gradient of non-perturbed plasma density is parallel to the electric field. This turbulence may result from a cascade of gradient drift instabilities, in which the primary instability produces growing plane waves that induce the instability of secondary perturbations, which in turn perturb the one-dimensionality of the wave field. The reverse action of these secondary perturbations may greatly enhance the self-action of the primary waves and energy transfer to the range of short-wavelength damped waves. Since drift waves obey a linear dispersion law, this cascade can be expected to result in generating strong turbulence. Note that this mechanism of energy transfer across the spectrum has a certain similarity to symmetry breaking in field theory. The system favors breaking the initial symmetry via instability, after which conditions are created in it for the most rapid energy transfer to the dissipation range. In addition to these fundamental arguments, our interest in drift turbulence is also supported by its widespread occurrence in ionospheric and laboratory plasmas and also by the important role that drift waves play in transfer processes.

Section 3 treats small-scale turbulence in hydrodynamic flows as the primary perturbation and analyzes the conditions under which it can cause secondary instability of the average large-scale wave motions in these flows. We consider this approach to be highly promising since large-scale perturbations in the atmosphere and the oceans in real geophysical conditions often develop not in laminar flows but in flows in which a certain level of small-scale turbulence has already set in; this turbulence greatly intensifies dissipative processes in comparison with the molecular transfer. The parameters of the small-scale turbulence do not remain constant in the process of evolution of large-scale perturbations, and dissipation intensity also changes with the parameters. Therefore, the stability conditions of geophysical flows may differ essentially from the stability conditions for laminar flows described by the same parameters. The main attention is typically paid in the literature to the evolution of weak wave perturbations against the background of fixed turbulence, or to the turbulence dynamics in the field of fixed wave perturbations, that is, one usually deals with problems in which the effect of one of these two components on the other is much larger than the reverse effect. In contrast, we will be looking for approaches to analyzing the cases in which the effect of the regular component on the random one and the effect in the opposite direction are of comparable magnitude.

4.2 Drift instability cascade and energy transfer in ionospheric turbulence

This section extends the concept of secondary instability to the case in which the system to be analyzed does not have multiparametric inequilibrium. In contrast to the examples given in the first two chapters, where the primary and secondary perturbations are plasma waves of various types, we continue here an analysis of secondary instabilities, mostly following Moiseev *et al* (1981), in a situation in which secondary perturbations are waves of the same type as the primary ones. In this case the difference between the two is that the wave vector of the primary perturbation falls within the region of linear instability while the wave vectors of the secondary perturbations, perpendicular to this vector, do not fall within this region. For this reason, the secondary perturbations remain stable until the primary ones grow to a certain threshold level.

Sudan *et al* (1973) proposed to use a process of this type to explain the generation of experimentally observable small-scale perturbations of ionospheric plasma density in the equatorial electrojet at altitudes of about 100 km. The characteristic size of these inhomogeneities is several meters while the minimum scale of perturbations that are unstable in the linear approximation is 30 to 60 m in conditions that are typical of this region of ionosphere during observations. The observed perturbations thus cannot be excited as a direct result of linear instability (which we will call the primary one) but can only arise as secondary ones due to nonlinear processes at a certain stage of the evolution of the instability.

The source of instability in this case is the combination of three factors: vertical inhomogeneity of plasma density at the lower boundary of the E-level of the ionosphere, the electric field $\boldsymbol{E}$ parallel to the plasma density gradient, and the essential difference in the behavior of electrons and ions in this field, owing to the differences in the relation between the corresponding collision frequencies and the cyclotron frequencies. Following Farley and Balsley (1973), Sudan *et al* (1973) give the following values for the electron (Ω_e) and ion (Ω_i) cyclotron frequencies and effective collision frequencies of electrons (ν_e) and ions (ν_i) at a specific region of the ionosphere (at an altitude of about 105 km): $\Omega_e = 5 \times 10^6\,\mathrm{c}^{-1}$, $\Omega_i = 90\,\mathrm{c}^{-1}$, $\nu_e = 4 \times 10^4\,\mathrm{c}^{-1}$, $\nu_i = 2.5 \times 10^3\,\mathrm{c}^{-1}$ (the last two are mostly determined by collisions with neutral particles).

It is easy to show that $\nu_e \ll \Omega_e$ and $\nu_i \gg \Omega_i$, that is, electrons are magnetized and are driven by the vertical electric and horizontal meridionally directed magnetic field $\boldsymbol{B}$ of the Earth along the magnetic equator, at a velocity $V_d = c\boldsymbol{E} \times \boldsymbol{B}/B^2$. As a frequently encountered value of drift velocity, Sudan *et al* (1973) give $V_d = 100\,\mathrm{m\,c}^{-1}$, thus identifying the corresponding strength of the non-perturbed electric field. At the same time, non-magnetized ions do not compensate for the drift electron current, so that non-zero electric current arises in the ionospheric plasma.

In homogeneous plasma, this current would remain stable until the drift velocity reached the threshold of the two-stream instability $V_{ts} = c_s(1 + \psi)$, in which $c_s = \sqrt{(T_e + T_i)/m_i}$ is the ion sound velocity, T_e and T_i are the electron and ion temperatures, respectively, m_i is the ion mass and $\psi = \nu_e\nu_i/(\Omega_e\Omega_i)$ (Buneman 1963, Farley 1963). Sudan *et al* (1973) give $c_s^2 = 10^5\ \mathrm{m}^2\ \mathrm{s}^{-1}$ for the characteristic value, which makes two-stream instability impossible at $V_d = 100\,\mathrm{m\,s}^{-1}$. At the same time, the values $V_d > c_s$ are also frequently encountered in the equatorial electrojet, causing plasma density perturbations of ion sound type, owing to the evolution and saturation of the two-stream instability (Farley 1985). These perturbations are known as perturbations of first type.

If, however, the density of the plasma is inhomogeneous and grows along the electric field (together with the potential energy of electrons in this field), then instability can also arise at $V_d < c_s$ but its nature is different: this is the drift gradient instability. This instability, which was discovered independently in experiments by Hoh (1963) and Simon (1963), is an analog of the well known interchange (flute) instability (Rosenbluth and Longmire 1957); the latter sets in when higher-density plasma is confined by lower-density plasma in the field of inertial forces. A number of authors followed Maeda *et al* (1963) and used this instability as a source of second-type perturbations of plasma density in the equatorial electrojet, observed at $V_d < c_s$; the model was gradually perfected by taking into account the effects of nonlinearity and inhomogeneity and achieving gradually more complete agreement of the results of this analysis to the data of remote and direct contact measurements (a review of the history and current status of this problem can be found in the publications of Farley 1985, Ronchi *et al* 1990 and Albert *et al* 1991).

Among the numerous models existing in the literature of nonlinear evolution of this instability, we regard the model presented by Sudan *et al* (1973) as the most attractive, since it allows, to a greater degree than the others, the application of the concept of secondary instabilities discussed in this book. Actually, the model requires considerable improvement: it does not take into account the effect of spatial inhomogeneity of the drift velocity (induced by the primary perturbation) on the dynamics of the secondary perturbations. At the same time, this effect must be taken into account since a shift in the drift velocity can suppress the drift gradient instability (as shown later by Huba and Lee 1983). For this reason, we will analyze below the dynamics of secondary perturbations within the framework of the model and take this effect into account and then, as suggested by Moiseev *et al* (1981), will analyze the feedback to the primary perturbations. On the other hand, we are interested only in the case $V_d < c_s$, in which the effect of ion inertia on perturbation dynamics at frequencies $\omega \ll \nu_i$ is not significant; the model can thus be somewhat simplified by ignoring the inertia of ions.

Therefore, following Moiseev *et al* (1981), we now consider the dynamics of electrostatic quasi-neutral perturbations at wavelengths much shorter than the Debye and Larmor electron radii; these perturbations propagate in the plasma with isothermal singly-charged ions and electrons strictly at right angles to the magnetic field. To describe the dynamics, we make use of the appropriately simplified set of equations of the two-fluid hydrodynamics of the plasma,

$$\frac{e}{m_e}\left(\frac{[\boldsymbol{V}_e\boldsymbol{B}]}{c} - \nabla\Phi\right) + \frac{T_e}{m_e}\frac{\nabla n_p}{n_p} + \nu_e\boldsymbol{V}_e = 0 \tag{4.2.1}$$

$$\frac{e}{m_i}\nabla\Phi + \frac{T_i}{m_i}\frac{\nabla n_p}{n_p} + \nu_i\boldsymbol{V}_i = 0 \tag{4.2.2}$$

$$\frac{\partial n_p}{\partial t} + \nabla(n_p\boldsymbol{V}_e) = 0 \tag{4.2.3}$$

$$\frac{\partial n_p}{\partial t} + \nabla(n_p\boldsymbol{V}_i) = 0 \tag{4.2.4}$$

where $n_p \equiv n_e \equiv n_i$ is the plasma density that stands for the simultaneously equal ion and electron concentrations, Φ is the electric field potential, V_e and V_i are the macroscopic velocities of motion of the electron and ion components of the plasma, and e and m_e are the magnitude of the electron charge and electron mass, respectively.

Note that these equations do not describe the stationary state of the plasma which forms as a result of a balance of the processes of ionization and recombination ignored in equations (4.2.3) and (4.2.4), and also of the horizontal stationary component of the electric field, which compensates for the vertical ion current by the vertical electron drift. However, the vertical velocities of electrons and ions, and also the correction to the velocity of the horizontal drift of electrons due to the gradient of the partial pressure of non-perturbed plasma, are small in comparison with V_d under conditions typical of the equatorial electrojet, so we can ignore them. Furthermore, we will not consider the effects connected with the dependence of the drift velocity V_d on altitude, the characteristic scale of inhomogeneity $L = n_0/|\nabla n_0|$ in non-perturbed plasma and collision frequencies ν_e and ν_i, since these effects were fairly thoroughly investigated for conditions in the ionosphere (Ronchi *et al* 1990); we focus our attention on the dynamics of secondary instability. When necessary, we will, without explicitly mentioning it, neglect small corrections of the order of ν_e/Ω_e, ν_i/Ω_i and the ratio of the perturbation scale to L. For L, Sudan *et al* (1973), following Farley and Balsley (1973), give a characteristic value $L \cong 6$ km.

We choose the right-handed Cartesian frame of reference, directing the coordinate axis x along the magnetic field and the z-axis along the gradient of non-perturbed plasma density (for the ionosphere this means northward and upward, respectively, and the y-axis then points westward). We are interested in two-dimensional perturbations that propagate strictly at right angles to magnetic field (since they are least subject to dissipation attenuation) and so consider the set of equations (4.2.1)–(4.2.4) in the plane yz. To describe perturbations, we introduce the quantities $n = n_p - n_0$ and $W = [T_e \ln(n_p/n_0) - e(\Phi - \Phi_0)]/m_e$. In view of the above remarks on small corrections, we now exclude V_e and V_i and obtain a nonlinear set of two equations for these quantities:

$$(1+\psi)\frac{\partial n}{\partial z} + V_d\frac{\partial n}{\partial y} + \frac{1}{\Omega_e}\left(\frac{\partial n}{\partial y}\frac{\partial W}{\partial z} - \frac{n_0}{L}\frac{\partial W}{\partial y} - \frac{\partial n}{\partial z}\frac{\partial W}{\partial y}\right)$$
$$= \frac{\psi}{\nu_i}c_s^2\Delta n \qquad (4.2.5)$$

$$\frac{\partial n}{\partial t} - \frac{c_s^2}{V_i}\Delta n = -\frac{\Omega_i}{\nu_i}\frac{n_0}{\Omega_e}\Delta W. \qquad (4.2.6)$$

If the nonlinear terms and the gradient drift are omitted, this set reduces to a single equation for n,

$$\left[\Delta + k_*\frac{\partial}{\partial y}\right]\frac{\partial n}{\partial z} + V_0\frac{\partial \Delta n}{\partial y} = D\Delta^2 n \qquad (4.2.7)$$

where

$$k_* = \frac{\nu_i}{\Omega_i L(1+\psi)} \qquad V_0 = \frac{V_d}{1+\psi},$$

and

$$D = \frac{c_s \nu_e^2}{\Omega_i \Omega_e (1+\psi)}$$

is the plasma diffusion coefficient in the direction at right angles to the magnetic field. Equation (4.2.7) for harmonic perturbations proportional to $\exp[\mathrm{i}(k_y + k_z - \omega t)]$ gives the dispersion equation

$$\omega = \omega_r + \mathrm{i}\gamma = \frac{k_y V_0 - \mathrm{i}Dk^2}{1 - \mathrm{i}(k_y k_* / k^2)} \tag{4.2.8}$$

where $k^2 = k_y^2 + k_z^2$. Separating the real and imaginary parts in (4.2.8) and neglecting the similar terms small in comparison with the terms retained, we obtain

$$\omega_r = \frac{k_y V_0}{1 + \left(k_* k_y / k^2\right)^2} \tag{4.2.9}$$

$$\gamma = \frac{\gamma_0 \left(\cos^2\theta - k^2/k_0^2\right)}{1 + \left(k_* k_y / k^2\right)^2} \tag{4.2.10}$$

where θ is the angle between the wave vector k and the (horizontal) y-axis, and

$$\begin{aligned} \gamma_0 &= \frac{V_i V_d}{\Omega_i L(1+\psi)^2} \\ k_0^2 &= \frac{\gamma_0}{D}. \end{aligned} \tag{4.2.11}$$

If the perturbation wave vector falls within the interval $k < k_0|\cos\theta|$ then $\gamma > 0$ and the perturbation grows with time. The minimum wavelength allowed for growing perturbations, $\lambda_m = 2\pi/k_0$, is reached for $\theta = 0$, that is, for horizontally propagating waves (moving along y). For the parameters of the ionospheric plasma in the equatorial electrojet as given above, $\lambda_m \simeq 30$ m (Sudan *et al* 1973). The expression for the instability increment in the interval $\lambda < \lambda_* = 2\pi/k_*$ was first derived by Rogister and D'Angelo (1970). For the parameters of the electrojet assumed above, $\lambda_* = 1.66$ km. The maximum instability increment is also reached at $\theta = 0$,

$$k_f = (k_*^2 + k_*(k_0^2 - k_*^2)^{1/2})^{1/2} \cong \sqrt{k_0 k_*}.$$

The wavelength corresponding to it is $\lambda_f = 2\pi/k_f \cong \sqrt{\lambda_0\lambda_*}$, which in this case comes to about 220 m, that is, about $7\lambda_m \ll L$.

At the values of the parameters given above, plasma density perturbations are experimentally observed at wavelengths of the order of several meters, that is, they are definitely stable in the linear approximation; to explain their generation, we need to take into account the nonlinear effects that are produced when the amplitudes of the most unstable wave grow to a considerable level. It is not difficult to see from the expression (4.2.11) for the real part of the drift wave frequency that maximum-increment waves are in fact non-dispersive since their phase velocity is almost independent of wave number.

In a large number of cases, the effect of instability on non-dispersive waves of various physical nature is front sharpening, up to formation of shock waves, and the ensuing transfer of energy from larger to smaller scales (Whitham 1974). However, the structure of nonlinear terms in equation (4.2.5) is such in our case that a one-dimensional harmonic wave (and also any sum of such waves with parallel wave vectors) is the exact solution of the set (4.2.5)–(4.2.6). Therefore, a wave with maximum increment can reach high amplitudes at unaltered phase velocity and without profile distortion (Sudan *et al* 1973).

The only way to open a channel for energy transfer to smaller scales is the spontaneous breaking of one-dimensional symmetry, which may be caused by secondary instability of small-scale perturbations having vertical wave vector in the field of the primary wave that induces horizontal gradients of plasma density and the vertical component of drift velocity. Since the mutual orientation of these two vectors is important for the evolution of the secondary instability, we follow Sudan *et al* (1973) and write the solution of equation (4.2.7), corresponding to the primary wave, in the form

$$n_1 = n_0 A_1 \sin\xi \qquad \xi = k_1 y - \omega_1 t.$$

Then the horizontal component of plasma density is given by

$$\frac{\partial n_1}{\partial y} = n_0 k_1 A_1 \cos\xi = k_1 L A_1 \frac{\partial n_0}{\partial z} \cos\xi.$$

Assuming $k_1 \sim k_f \gg 1/L$, we find that even small amplitudes of the primary wave A_1 of the order of several per cent are sufficient for this component to greatly exceed the stationary vertical component $\partial n_0/\partial z$. The vertical component of the electron drift velocity induced by the primary wave,

$$V_{ze} = -\frac{\nu_i V_d A_1 \sin\xi}{\Omega_i(1+\psi)} \tag{4.2.12}$$

may become, at the same amplitude A_1, comparable with V_d because $\nu_i/\Omega_i \gg 1$. The product $V_{ze}\partial n_1/\partial y$ can be used, by analogy to (4.2.10), to evaluate the local growth rate of the expected secondary instability with respect to the perturbations with vertical wave vector whose modulus is much greater than k_1.

Arguments of this type indicate that this product may grow much larger than V_d/L even at relatively small values of A_1. In this case we can expect that the local increment of the secondary instability will increase to become much greater than that of the primary instability, and, by analogy to (4.2.11), the instability boundary will significantly shift to shorter wavelengths. Now we can neglect changes in the amplitude of the primary wave over the time of growth of the secondary instability, except for the cases in which the secondary instabilities are localized in the vicinity of the phases $\xi_n = n\pi/2$, where n is an arbitrary integer. If $\xi = \xi_n$, the product $V_{ze}\partial n_1/\partial y = 0$ and the evolution of secondary perturbations close to this phase remains slow at any amplitude of the primary wave. Beyond these neighborhoods, the secondary instability can be studied using the equations in which the parameters of the primary wave are time-independent and their space scale is large compared with the secondary wave wavelength; this makes it possible to obtain from (4.2.5)–(4.2.6) for small secondary perturbations $|n_2| \ll |n_1|$ one equation, linearized in n_2, in the reference frame moving together with the primary wave:

$$\left[\Delta - \frac{\nu_i}{\Omega_i n_0(1+\psi)} \frac{\partial n_1}{\partial y} \frac{\partial}{\partial z}\right] \frac{\partial n_2}{\partial t} + \frac{V_{ze}}{(1+\psi)} \Delta \frac{\partial n_2}{\partial z} = D\Delta^2 n_2. \quad (4.2.13)$$

In contrast to the similar equation (4.2.7) that describes the primary instability, the coefficients $\partial n_1/\partial y$ and V_{ze} in equation (4.2.13) are not constant but depend on the y-coordinate. Since these coefficients remain constant along the z-coordinate, we can see the solution of equation (4.2.13) as a harmonic function of z:

$$n_2 = \mathrm{Re}\left[a(y,t)\,\mathrm{e}^{\mathrm{i}k_z z}\right].$$

Equation (4.2.12) then gives us an equation for the complex amplitude $a(y,t)$:

$$\left(\frac{\partial^2}{\partial y^2} - k_z^2 - \mathrm{i}\frac{\nu_i k_z}{\Omega_i(1+\psi)n_0} \frac{\partial n_1}{\partial y}\right) \frac{\partial a}{\partial z} + \mathrm{i}\frac{V_{ze}k_z}{(1+\psi)} \left(\frac{\partial^2}{\partial y^2} - k_z^2\right) a$$
$$= D\left(\frac{\partial^2}{\partial y^2} - k_z^2\right)^2 a. \quad (4.2.14)$$

The y-dependent coefficients $\partial n_1/\partial y$ and V_{ze} enter this equation with weights that are proportional to different powers of k_z. This gives us sufficient grounds to expect that, depending on the value of k_z, either one or the other inhomogeneity will dominate. Having focused our attention on the evolution of short-wave secondary perturbations already in deriving equation (4.2.13), we will consider the case of k_z sufficiently large for the predominance of the shift V_{ze}. In this case the coefficient $\partial n_1/\partial y$ in the neighborhood of a phase ξ_* of the primary wave can be regarded constant,

$$\frac{\partial n_1}{\partial y}(\xi) = \frac{\partial n_1}{\partial y}(\xi_*)$$

and V_{ze} can be rewritten as

$$V_{ze}(\xi) = (1+\psi)V_p(y) = (1+\psi)[V_* + V_p'(y - y_*)].$$

For equation (4.2.14) we can now write the explicit form of the approximate solution:

$$a = a_0 \exp\left\{\mathrm{i}k_z[z - V_p(y)t] + \frac{\gamma_*}{V_p'}\arctan(V_p' t) - Dk_z^2 t\left[1 + (V_p' t)^2\right]\right\} \quad (4.2.15)$$

where

$$\gamma_* = -\frac{\Omega_i V_{ze}(\xi_*)}{\nu_i(1+\psi)^2 n_0}\frac{\partial n_1}{\partial y}(\xi_*) = \frac{V_{ze}(\xi_*)}{V_d}V_p' = \frac{k_1 V_d V_1^2 \sin 2\xi_*}{2(1+\psi)}$$
$$V_1 = \frac{\nu_i A_1}{\Omega_i(1+\psi)}. \tag{4.2.16}$$

This solution ignores the long-wavelength corrections to the dispersion of secondary waves in the field of the primary wave; these corrections are similar to the appropriate corrections to the dispersion of primary waves represented by the denominators in the expressions (4.2.9) and (4.2.10), since it is assumed that k_z is sufficiently large for these corrections to be negligibly small. The solution (4.2.15) demonstrates the effect of the shift of the induced drift velocity V_{ze}: since this shift slants the wave crests, the distances between them decrease and they fall into the range of diffusion damping even if $\gamma_* > Dk_z^2$ and perturbations could grow at the initial stage. If $\gamma_* \gg Dk_z^2$, the bounds on the time range of the initial growth of perturbations can be given approximately by the relation

$$|V't| < \left(\frac{\gamma_*}{Dk_z^2}\right)^{1/4} \equiv \tau(\xi_*).$$

In this case the secondary wave amplitude will increase over the entire period of growth by a factor

$$\frac{a_{\max}}{a_{\min}} = \exp\left(2\frac{\gamma_*}{|V_p'|}\arctan\tau\right). \tag{4.2.17}$$

Let us consider how this ratio depends on the primary wave phase ξ. Using now (4.2.17) and (4.2.12), we can write

$$\frac{\gamma_*}{|V_p'|} = V_1|\sin\xi_*| \tag{4.2.18}$$

$$\tau = \left(\left(\frac{k_1 k_0^2 V_1^2 \sin 2\xi_*}{2(1+\psi)k_z^2 k_*}\right)^{1/2} - 1\right)^{1/2}. \tag{4.2.19}$$

Substituting (4.2.18) and (4.2.19) into (4.2.17), we find that the maximum growth of secondary waves is achieved close to the primary wave phase, $\xi_* \cong 2\pi/5+n\pi$, where n is an arbitrary integer. The maximum of the exponential in (4.2.17) grows with increasing V_1 and decreasing k_z. We have pointed out already that as k_z decreases, the influence of the non-uniform plasma density gradient in the primary wave is enhanced. In this situation its effect is seen in the growing importance of the inhomogeneity of the amplitude of the growing waves along y, so that the (4.2.15)-type solution requires modification; indeed, it was written under the assumption that the derivatives with respect to y in (4.2.14) are mostly determined by the product $ik_z V_p(y)t$ in the exponent of the exponential of (4.2.17) and that the contribution of the derivatives of the second term in this exponent is small. As k_z decreases, this assumption becomes invalid, the derivatives of the second term in the exponent of the exponential of (4.2.17) become large and result in the slowdown and then termination of the growth of the amplification coefficient of secondary waves in the field of the primary wave at k_z values $\cong 4k_1$, after which the amplification coefficient starts to decrease again. The maximum amplification coefficient then corresponds to the exponent of the exponential in (4.2.17) quite close to $2V_1$, that is, the initial perturbations will grow at $V_1 \cong 3$ by a factor of several hundred.

An analysis of the secondary instability, linear in the amplitude of secondary perturbations, shows that the shift in the drift electron velocity induced by the primary wave suppresses the unlimited growth of secondary perturbations but allows their amplification from the initial level by a finite factor. Assuming now that the initial level of secondary perturbations is independent of the phase of the primary wave, we can expect the maximum amplitude of the amplified secondary perturbations to coincide with the maximum of the amplification coefficient. Its position is then shifted relative to the maximum of the local increment of the secondary instability $\gamma_*(\xi_*)$ towards lower values of the shift of the drift velocity $V_p'(\xi_*)$.

After the maximum amplitude has been reached, the secondary perturbations keep being tilted by the shift of the drift velocity in the primary wave and are also damped out by diffusion. Before their amplitude decreases substantially, their wave number may grow up to (2 to 3)k_0. Therefore, high-intensity short-wave perturbations, not generated by the primary instability, may arise in response to the secondary perturbation.

Since secondary perturbations may reach considerably high amplitudes in the vicinity of their maximum, we may expect that the dynamic feedback to the primary wave may also prove to be significant.

In order to evaluate its effect on the dynamics of the primary instability, we substitute n and W into the set of equations (4.2.5) and (4.2.6) as a sum of the primary and secondary waves: $n = n_1 + n_2$, $W = W_1 + W_2$. Since the primary wave propagates horizontally, that is, n_1 and W_1 are independent of z, and undergoes no linear self-action, the set (4.2.5)–(4.2.6) can be reduced to a single equation for n_1 in the reference frame that moves at the phase velocity of the primary wave,

$$\frac{\partial n_1}{\partial z} = \gamma_0 n_1 + D\Delta n_1 - \frac{1}{\Omega_e(1+\psi)}\left\langle \frac{\partial n_2}{\partial y}\frac{\partial W_2}{\partial z} - \frac{\partial n_2}{\partial z}\frac{\partial W_2}{\partial y} \right\rangle \tag{4.2.20}$$

where $\langle\ldots\rangle$ stands for averaging over the period of the secondary wave. To evaluate the nonlinear term, it is necessary to reconstruct the relation between W_2 and n_2 that we used in deriving (4.2.13). This relation takes the simplest form for short-wavelength secondary waves for which we need not take into account the spatial inhomogeneity of the plasma density gradient in the primary wave. We can now write

$$\frac{W_2}{\Omega_e} = -\frac{\nu_i V_{ze}}{\Omega_i k_z^2(1+\tau^2)n_0}\frac{\partial n_2}{\partial z} = \frac{\nu_i^2 V_d n_1}{\Omega_i^2(1+\psi)k_z^2(1+\tau^2)n_0}\frac{\partial n_2}{\partial z}. \qquad (4.2.21)$$

Substituting (4.2.21) into (4.2.20) and denoting the maximum value of the secondary wave amplitude by $A_2(y)$, we rewrite (4.2.20) for each phase of the primary wave phase in the form

$$\frac{\partial n_1}{\partial z} = \gamma_0 n_1 + D\Delta n_1 + \frac{\nu_i^2 V_d}{2\Omega_i^2(1+\psi)^2(1+\tau^2)}\frac{\partial}{\partial y}(n_1 A_2^2).$$

Analyzing this equation, we immediately notice that already at relatively low values of A_2 of the order of $\Omega_i(1+\psi)/\nu_i \cong 0.04$ the growth of n_1 stops in the neighborhood of the maxima and minima of the primary wave but, on the other hand, increases at maxima and minima of the amplitude of A_2 along the direction of propagation of the primary wave, that is, the crests and troughs of the primary wave become flatter while the slopes get steeper, and the wave shape tends to rectangular. This shape of large-scale drift waves is indeed observed in the region of the equatorial electrojet (Pfaff *et al* 1987). Also suppressed, in addition to the density gradient near crests and troughs of the primary wave, is the shift of the wave-induced vertical component of the drift velocity, which stabilizes the secondary wave amplitude. In their turn, these secondary waves produce vertical gradients of plasma density, greatly exceeding the initial gradient; this makes possible the generation of horizontally propagating waves with wave vectors that go far beyond the region of the primary instability. Such waves are regularly observed in experiments (Sudan *et al* 1973, Farley and Balsley 1973, Farley 1985). Therefore, a cascade of drift instabilities is generated owing to spontaneous breaking of the initial symmetry, and energy is channeled via the spatial spectrum from the region of primary instability to the region of diffusion damping.

4.3 Wave-turbulent instability

We will now analyze the conditions of evolution of secondary instabilities in situations in which the primary perturbations are turbulent pulsations. In this case the secondary instability is caused by the interaction of the three components of the hydrodynamic flow (mean flow, regular wave, stochastic motions). The stochastic component draws energy from the mean flow and feeds it into the regular wave; in its turn, the regular wave increases the efficiency of interaction between stochastic motion and the other two components. If the amplitude of the regular wave is small, this process can be treated in the framework of perturbation theory.

In the zero approximation one considers the mean flow that carries the regular harmonic wave and stochastic motions. The small amplitude of the wave allows one to ignore its self-action and the interaction with other waves.

The next approximation considers small distortions that are introduced into the flow by the regular wave and takes into account the distortion-induced changes in the conditions of existence of stochastic motions; these changes result in a modulation of the field of stochastic motions. The interaction of the modulated field of stochastic motions with the wave is a resonance process and must be taken into account as a slow change in the amplitude and phase of the wave.

First we follow the work of Kolykhalov *et al* (1989) and analyze the interaction of acoustic waves with turbulence in undersonic plane-parallel flows with smooth velocity profile. The acoustic wave may change its amplitude owing to the generation of an in-phase wave by the spatially modulated turbulence. In its turn, if a statistical inhomogeneity is present in the non-perturbed field of turbulent pulsations, the spatial modulation of turbulence by the acoustic wave may be achieved in a purely mechanical way (by convective transfer); influence of generation, dissipation and diffusion of turbulence, whose characteristic times are large in comparison with the period of the acoustic wave is insignificant.

Let a fluid with unit non-perturbed density and constant sound velocity c_s be driven in a steady-state manner along the x-axis at an average velocity $U \ll c_s$ (U is a function of the coordinate y) through an H-thick layer bounded by sound-reflecting planes, by a pressure gradient $\mathrm{d}P_0/\mathrm{d}x$ and a single non-zero component of the tensor of turbulent tangential stress $\tau_0(y) \equiv -2\langle u_x u_y \rangle$, where u_i are the turbulent pulsations of velocity.

The acoustic wave is, in the zero approximation, a standing wave along y and a running wave along x, and is described by pressure oscillations $P = P_a \cos(k_y y)\cos\varphi$, where $\varphi = \omega t - k_x x$, $\omega^2 = k^2 c_s^2$, $k^2 = k_x^2 + k_y^2$, and oscillations of velocity components $\{V_x, V_y\} = P_a\{k_x \cos(k_y y), k_y \sin(k_y y)\}\cos\varphi/\omega$. In the first approximation, an acoustic wave emitted by the modulated turbulence is added to the acoustic wave above; the two are in phase. Pressure oscillations in the second wave can be found from the equation (Lighthill 1952)

$$\frac{1}{c_s^2}\frac{\partial^2 P_1}{\partial t^2} - \Delta P_1 = -\frac{\partial^2 \tau_1}{\partial x\,\partial y} \tag{4.3.1}$$

where τ_1 is a zero-approximation correction term to τ_0 due to the acoustic wave. We will find it from the transfer equation of turbulent stress whose left-hand side is obtained by direct averaging of the equations of motion for turbulent velocity pulsations, while the right-hand side is assumed to be zero on the basis of arguments on the smallness of the period of the acoustic wave in comparison with the characteristic time of turbulence evolution in the flow,

$$\frac{\partial \tau_1}{\partial t} + V_y \frac{\partial \tau_0}{\partial y} = 0.$$

Therefore,

$$\tau_1 = \frac{k_y P_a}{\omega^2}\frac{\mathrm{d}\tau_0}{\mathrm{d}y}\sin(ky)\cos\varphi.$$

For the correction P_1 to remain limited, it is necessary to modify (4.3.1) by assuming a slow dependence $P_a(t)$ given by the equation

$$\frac{1}{P_a}\frac{\mathrm{d}P_a}{\mathrm{d}t} = 2\frac{k_x k_y^2}{\omega k^2 L}\int_0^L \mathrm{d}y\,\sin^2(k_y y)\,\frac{\mathrm{d}\tau}{\mathrm{d}y}. \tag{4.3.2}$$

The right-hand side of this expression is the growth rate of the acoustic wave, which is positive for the wave propagating in the direction of reduction of the mean pressure.

Together with this process, turbulence scatters the sound, which reduces the amplitude of the primary wave with a decay rate $\delta \propto k^2 b_0 L/c_s$, where $b_0 = \langle u_i u_i \rangle$ and L is the macroscopic scale of turbulence. Using the widely used Kolmogorov scheme (Kolmogorov 1942), the same macroscopic scale L can be used to evaluate the derivative $\mathrm{d}\tau_0/\mathrm{d}y$ in (4.3.2), assuming its value to be of the order of $L\sqrt{b_0}\,U_{\max}/H^2$. Comparing then the value of δ with the obtained growth increment of the acoustic wave, we readily see that scattering does not significantly affect the secondary instability as long as

$$k \ll \frac{1}{H}\left(\frac{U_{\max}^2}{b_0}\right)^{1/4}.$$

Therefore, acoustic waves of not too short wavelength must increase with increasing depth of modulation of turbulence.

We conclude that the presence of inhomogeneity of tangent stress, caused by the evolution and saturation of the primary instability that generates turbulence, leads to the onset of secondary instability that tends to involve in the relaxation process still newer degrees of freedom connected with acoustic waves.

Now we will analyze a more complicated situation in which the period of a regular wave is sufficiently large for the wave to be able to modify the conditions of generation and suppression of turbulence.

Following Moiseev *et al* (1982, 1984) and Moiseev and Pungin (1996), we consider the propagation of a large-scale finite-amplitude internal wave in stratified shear flow that consists of one or several layers whose parameters are close to the turbulence generation threshold. The internal wave of small but finite amplitude forces the parameters of these layers to periodically cross the threshold, which produces a primary instability in the layers and generates turbulence. This results in a significant spatially inhomogeneous component of turbulence energy b and, correspondingly, variable coefficients of turbulent transfer. This produces turbulent fluxes of momentum and buoyancy, in-phase with the wave and proportional to the constant gradients of the relevant quantities in non-perturbed flow; these can be directed against the fluxes generated by the spatially uniform component of turbulence and may exceed these fluxes in magnitude. In this case the wave is not damped by turbulent viscosity but, in contrast, is enhanced at the expense of the energy of the mean flow.

The idea of this wave-turbulent instability (WTI) was first formulated by Chimonas (1972); however, some aspects of his work needed reconsideration, for example, the instability condition that he proposed. Numerical simulation of WTI (Fua *et al* 1982) confirmed that this instability is possible but failed to identify the condition under which it could be observed.

This condition was analytically derived by Moiseev *et al* (1982, 1984) from the Euler equation for wave perturbations (averaged over turbulent pulsations) and the Kolmogorov-type equation for energy density of turbulence (Kolmogorov 1942), linearized over wave amplitude. However, the latest turbulence models of stratified fluid make possible a more rigorous description of the effect of stratification on the evolution of turbulence than the model studied in the above papers. Thus Moiseev and Pungin (1995, 1996) adapted the numerical model that was previously developed and successfully tested by Gibson and Launder (1976), and applied it to analytically scrutinize the conditions of evolution of turbulent wave instability in a physical system comprising three components:

- mean plane-parallel flow of fluid with statistically stable stratification of density $\rho_0(z)$ and vertically non-uniform distribution of the only (horizontal) velocity component $U(z)$ (z is the upward-increasing vertical coordinate);
- regular large-scale wave motions;
- small-scale turbulence.

The existence of stable density stratification that occurs for real values of the Brunt–Väisälä frequency (buoyancy frequency) N that is given by the expression (Gossard and Hook 1975)

$$N^2 = -\frac{g}{\rho_0}\frac{\mathrm{d}\rho_0}{\mathrm{d}z} - \frac{g^2}{c_s^2}$$

where g is the gravity acceleration and c_s is the sound velocity, permits the flow to remain laminar no matter how large the values of the Reynolds number, provided the vertical velocity shear $U' \equiv \mathrm{d}U/\mathrm{d}z$ is not sufficient for overcoming the stabilizing effect of stratification, while there are no horizontal shifts and no external forces capable of sustaining turbulence. The vertical velocity shift necessary for producing turbulence is found in terms of a dimensionless parameter $\mathrm{Ri} = N^2/U'^2$ known as the dynamic Richardson number (Turner 1973). If $\mathrm{Ri} > 1/4$, infinitesimal flow perturbations cannot grow exponentially, otherwise this constraint is cancelled (Miles 1961).

The threshold value of the Richardson number $\mathrm{Ri}_{\mathrm{cr}}$, which bounds from above the range of values at which pulsations can grow, may have a value close to $1/4$ for turbulent pulsations, since these constitute perturbations of small but finite amplitude. Note that the sign of deviation of $\mathrm{Ri}_{\mathrm{cr}}$ from $1/4$ may be arbitrary, depending on the type of influence of the nonlinearity. If the Richardson number is above this critical value over the entire flow, then the relaxation of the non-uniform velocity distribution is suppressed. If, however, we have a layer $z_0 - h/2 < z < z_0 + h/2$ where the stability reserves are low: $\mathrm{Ri}(z) = \mathrm{Ri}_{\mathrm{cr}} + \delta\mathrm{Ri}$, $0 < \delta\mathrm{Ri} \ll 1$, so that the large-scale internal wave of small but finite amplitude (sufficient to cause modulation of $\mathrm{Ri}(z_0)$ at an amplitude above $\delta\mathrm{Ri}$) can make possible a growth in turbulence in that part of its period when $\mathrm{Ri}(z_0)$ decreases and gets smaller than $\mathrm{Ri}_{\mathrm{cr}}$; the scale of the internal wave is large in comparison with the characteristic spatial and temporal scales of turbulence.

In the simplest scenario that makes it possible to avoid questions on the type of transition from the laminar to turbulent flow, we can assume that the system is subjected to a weak external force that sustains a certain minimal possible level of turbulence in the layer under investigation, in that part of the wave period when it causes $\mathrm{Ri}(z)$ to increase and thereby additionally stabilizes the flow. If the other part of the period, in which the wave destabilizes the flow, is such that the energy of turbulence has sufficient time to rise above this minimal possible level, then the reverse effect of turbulence on the wave will be mostly determined by its parameters in the destabilization phase of the wave. In this case the turbulence is a periodic sequence of spots that run along the layer of reduced stability together with the destabilizing parts of each wave period.

We find the simplest description of the interaction of the wave with the turbulence it induces in the case when the turbulence dynamics is dictated by the local values of wave fields and it is possible to ignore the resulting vertical inhomogeneity of its parameters. In order to find the conditions of applicability of this approximation, we first consider the turbulence dynamics in the spatially uniform case (when the flow parameters N and U' and hence Ri are independent of coordinates). Then we treat the specific features of relaxation of vertical inhomogeneity of turbulence parameters and of the flow in the near-threshold situation when the Richardson number over the entire flow differs only slightly from its critical value, $\mathrm{Ri} = \mathrm{Ri}_{\mathrm{cr}} + \delta\,\mathrm{Ri}(z)$, where $|\delta\,\mathrm{Ri}| \ll 1$; we do this using the model of description of turbulence (Gibson and Launder 1976) in which turbulence is characterized by two independent parameters: turbulence energy density b and its rate of dissipation ε. The main equations of this model for the case under discussion can be written as

$$\frac{\partial b}{\partial t} = \frac{\partial}{\partial z}\left(\nu_T \frac{\partial b}{\partial z}\right) + \nu_T (U')^2 (1 - \mathrm{Rf}) - \varepsilon \tag{4.3.3}$$

$$\frac{\partial \varepsilon}{\partial t} = \frac{\partial}{\partial z}\left(\frac{\nu_T}{\sigma_\varepsilon}\frac{\partial \varepsilon}{\partial z}\right) + c_{\varepsilon_1} \nu_T \frac{\varepsilon}{b}(U')^2 - c_{\varepsilon_2}\frac{\varepsilon^2}{b} \tag{4.3.4}$$

$$\frac{\partial U}{\partial t} = \frac{\partial (\nu_T U')}{\partial z} \tag{4.3.5}$$

$$\frac{\partial N^2}{\partial t} = \frac{\partial^2}{\partial z^2}\left(\frac{\nu_T}{\sigma_T} N^2\right). \tag{4.3.6}$$

Here σ_ε, c_{ε_1} and c_{ε_2} are the empirical constants of the model, ν_T is the turbulent viscosity, σ_T is the ratio of the coefficients of the turbulent viscosity and heat conductance (analogous to the Prandtl number), and $\mathrm{Rf} = \mathrm{Ri}/\sigma_T$ is the dynamic Richardson number. The best fit of the results of numerical calculations to

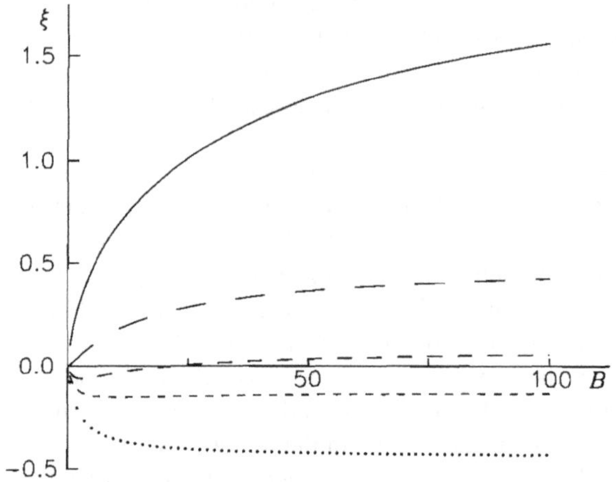

Figure 4.1. Graphs of $\xi(B)$ for various values of the Richardson number Ri: ——, Ri = 0.5; — — —, Ri = 1.0; - - -, Ri = 1.5; - - - -, Ri = 2.0; ······, Ri = 4.0.

experimental data was achieved by the authors of the model for the following choice of constants (Gibson and Launder 1976): $\sigma_\varepsilon = 1.3$, $c_{\varepsilon_1} = 1.45$, $c_{\varepsilon_2} = 1.9$.

To calculate ν_T and σ_T, the model uses algebraic relations which, in their original form (Gibson and Launder 1976), allow us to close the set of equations (4.3.3)–(4.3.6) at every step of numerical calculations. Using these calculations, we can analytically derive an algebraic equation to close the set (4.3.3)–(4.3.6) which relates the parameters Ri, $\xi = \nu_T(U')^2(1 - \mathrm{Rf})/\varepsilon$ and $B \equiv N^2b^2/\varepsilon^2$ (Moiseev and Pungin 1995); the equation includes empirical constants of the model as its parameters. Graphs of $\xi(B)$ determined from this equation for several values of Ri are shown in figure 4.1.

In the spatially uniform case, when N, U', b and ε are independent of z, the set (4.3.3)–(4.3.6) reduces to a set of two autonomous ordinary differential equations for b and ε, since N and U' remain constant and affect the behavior of b and ε only parametrically. The shape of phase trajectories in the plane (b, ε) is determined by the value of a single parameter, namely the gradient Richardson number Ri, while the dimensional values of N or U' determine the time scale of motion along these phase trajectories.

For any non-negative value of Ri, the evolution of turbulence can be separated into two stages: the transient and the self-similar. The second stage determines the turbulence behavior in the limit $t \to \infty$. After the self-similar relation between b and ε reaches stationary stage at the second state, at small Ri, the turbulence energy grows exponentially, even though it may decrease at the first stage if the chosen initial value of ε is too large; this is in complete agreement with the limiting case of non-stratified fluid (Ri = 0) (Barenblatt

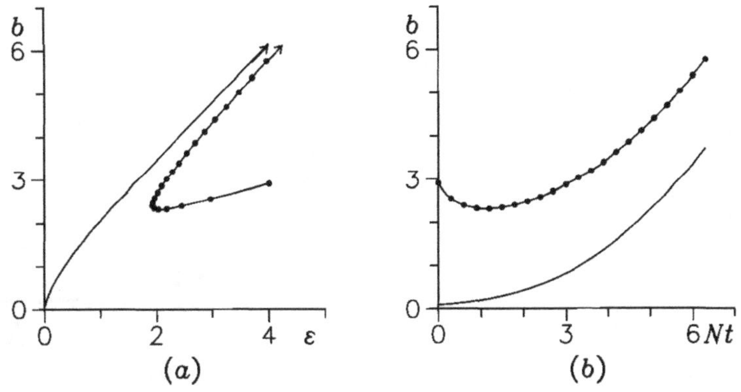

Figure 4.2. Phase trajectories in the $(b\varepsilon)$ plane (a) and time dependence of turbulence energy $b(t)$ (b) for $\mathrm{Ri} = 0.1 < \mathrm{Ri}_{\mathrm{cr}}$. The arrows indicate the direction of motion of the system along the phase trajectory.

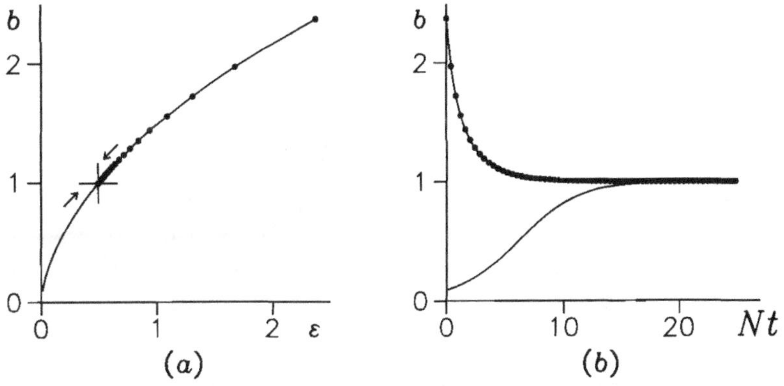

Figure 4.3. The same as figure 4.2 for $\mathrm{Ri} = \mathrm{Ri}_{\mathrm{cr}} \approx 0.225$. The + symbol indicates the steady state to which the system tends in its motion along the plotted phase trajectories.

1990). Typical examples of phase trajectories and time variations of the turbulent energy for this case are shown in figure 4.2.

As Ri increases, the turbulence growth rate at the second stage decreases and vanishes when the critical value $\mathrm{Ri} = \mathrm{Ri}_{\mathrm{cr}}$ is reached; this is dictated by the choice of the empirical constants of the model. If the constants are chosen as recommended by Gibson and Launder (1976), then $\mathrm{Ri}_{\mathrm{cr}} \approx 0.225$, that is, slightly less than $1/4$. If $\mathrm{Ri} = \mathrm{Ri}_{\mathrm{cr}}$, b and ε over the self-similar stage tend to certain limiting values that depend on the initial conditions. If the ratio of the initial

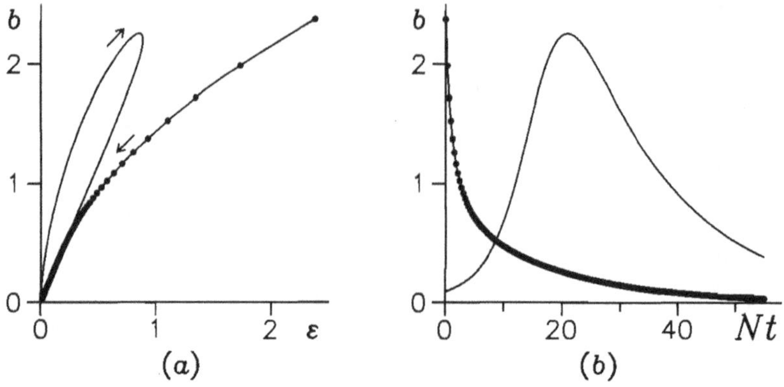

Figure 4.4. The same as figure 4.2 for Ri = 0.3 > $\mathrm{Ri_{cr}}$.

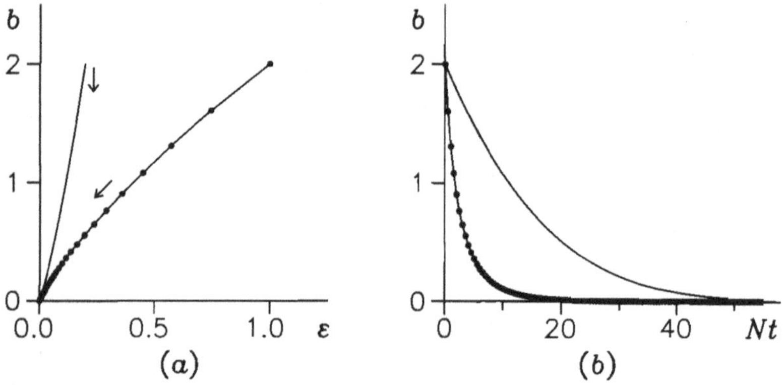

Figure 4.5. The same as figure 4.2 for Ri = 1 > $\mathrm{Ri_{cr}}$.

values of b and ε equals the self-similar value, then these two parameters remain equal to their initial values. The corresponding examples of phase trajectories and $b(t)$ curves are shown in figure 4.3.

As Ri grows larger than $\mathrm{Ri_{cr}}$, the turbulence energy at the self-similar stage tends to zero and we can identify three modes of damping that follow one another as Ri increases. These modes are also distinguished by two specific values by the Richardson number, $\mathrm{Ri_1}$ and $\mathrm{Ri_2}$, whose values also depend on the choice of constants of the model. With the choice as given above, we have $\mathrm{Ri_1} \approx 0.745$ and $\mathrm{Ri_2} \approx 1.8$. If $\mathrm{Ri_{cr}} < \mathrm{Ri} < \mathrm{Ri_1}$, then the turbulence energy may grow at the transient stage at sufficiently small initial values of ε and reach self-similar decay stage after reaching the maximum value that is significantly larger then the initial one. Examples of phase trajectories and $b(t)$ curves for this case are shown in figure 4.4. If $\mathrm{Ri} > \mathrm{Ri}_1$, then the decay is monotone for arbitrary initial values of b and ε (see figure 4.5); if $\mathrm{Ri} > \mathrm{Ri}_2$, the exponential decay at the self-similarity stage is replaced by the power-law stage $b(t) \propto t^{\alpha(\mathrm{Ri})}$, where the exponent α is a non-monotone function of the Richardson number and assumes its minimum value $\alpha_{\min} \approx 17$ at $\mathrm{Ri} \approx 5$.

For the values of Ri such that the turbulence energy increases or decreases exponentially at the self-similarity stage, the dissipation rate changes in proportion to energy, with a proportionality coefficient that is a function of Ri. The macroscopic turbulence scale, given by the expression $L \equiv \sqrt{b^3/\varepsilon^2}$ and characterizing the size of the largest vortices, also increases or decreases exponentially together with the turbulence energy and its dissipation rate, in proportion to the square root of turbulence energy. If deviations of Ri from $\mathrm{Ri}_{\mathrm{cr}}$ are small, the characteristic time of this growth or decline is much greater than the duration of the transient stage $\tau^* \propto 1/N$. If, in addition, Ri varies with a characteristic time $\tau \gg \tau^*$, then the turbulence evolution will correspond to the self-similarity mode for the current value of Ri and be described by a single equation,

$$\frac{\mathrm{d}b}{\mathrm{d}t} = -\zeta b N(\mathrm{Ri} - \mathrm{Ri}_{\mathrm{cr}}) \tag{4.3.7}$$

where ζ is determined by the values of the constants of the model; with the choice of constants as given above, ζ is very close to 1.

Assuming that the changes $\delta\,\mathrm{Ri} \equiv \mathrm{Ri} - \mathrm{Ri}_{\mathrm{cr}}$ are caused by a wave perturbation at local frequency $\Omega \ll N$, that is, at a period $T \equiv 2\pi/\Omega \gg \tau^*$, we easily find from equation (4.3.7) that considerable changes in turbulence energy $b_{\max}/b_{\min} \propto \exp(|\delta\,\mathrm{Ri}|T/\tau^*) \gg 1$ can be produced even at a small but finite perturbation amplitude $\tau^*/T \ll |\delta\,\mathrm{Ri}| \ll 1$.

Consider now the relaxation dynamics of vertical inhomogeneity in the distribution of b and $\delta\,\mathrm{Ri}$ in order to find the conditions under which this relaxation does not result in significant deviations from the local turbulence dynamics in the field of perturbations $\delta\,\mathrm{Ri}$ induced by the wave (see above).

Vertically inhomogeneous perturbations of the Richardson number, that is, of at least one of the two parameters N or U', violate, according to equations (4.3.5) and (4.3.6), the stationarity of these parameters; this is caused by the generated divergence of turbulent mass or momentum fluxes. The divergence of these fluxes is caused both by the vertical inhomogeneity of flow parameters and by the vertical inhomogeneity, induced by them, of the turbulence transfer coefficients.

We can compare the relative effects of these two factors using the momentum flux as example:

$$\frac{\delta(\nu_T U')}{\nu_T U'} = \frac{\delta \nu_T}{\nu_T} + \frac{\delta U'}{U'}.$$

The algebraic relations of the model for calculating ν_T in the self-similarity mode can be reduced to the form

$$\nu_T = c_\nu(\mathrm{Ri})b/N \tag{4.3.8}$$

where the form of the function $c_\nu(\mathrm{Ri})$ is dictated by the chosen constants of the model. For the constants chosen as described above we obtain $c_* \equiv c_\nu(\mathrm{Ri}_{\mathrm{cr}}) \approx 0.15$. In view of equations (4.3.7) and (4.3.8) we obtain an estimate (Moiseev and Pungin 1995)

$$\frac{\delta\nu_T}{\nu_T} = 2\,\mathrm{Ri}_{\mathrm{cr}}\frac{\delta U'}{U'}\left(\zeta N\tau - \frac{\mathrm{d}F}{F\mathrm{d}\,\mathrm{Ri}}\right).$$

In the case under consideration, $NT \gg 1$, we find $\delta\nu_T/\nu_T \gg \delta U'/U'$. This means that perturbations of turbulent flows are mostly governed by the inhomogeneity of the turbulence transfer coefficients, not of wave fields, so that we can expect that the reverse effect of these fluxes is not reducible to conventional dissipation.

Retaining only the first of the components of turbulent fluxes mentioned above, and replacing equations (4.3.3) and (4.3.4) by the self-similar equation (4.3.7), we can derive a closed set of equations for the case of small deviations from the homogeneous stationary state that can be written in the form $\mathrm{Ri}(z,t) = \mathrm{Ri}_{\mathrm{cr}} + \delta\,\mathrm{Ri}(z,t)$, $b(z,t) = b_0 + \delta b(z,t)$, $|\delta\,\mathrm{Ri}|/\mathrm{Ri}_{\mathrm{cr}}$, $|\delta b|/b_0 \ll 1$ (which implies that for turbulent viscosity we can also write $\nu_T(z) = \nu_0 + \delta\nu(z,t)$, $|\delta\nu|/\nu_0 \ll 1$):

$$\begin{aligned}\frac{\partial\,\mathrm{Ri}}{\partial t} &= \left(\frac{1}{\sigma_T} - 2\right)\mathrm{Ri}_{\mathrm{cr}}\frac{\partial^2\nu_T}{\partial z^2}\\ \frac{\partial\nu_T}{\partial t} &= -\zeta\nu_0 N(\mathrm{Ri} - \mathrm{Ri}_{\mathrm{cr}}).\end{aligned} \tag{4.3.9}$$

This set describes the perturbation relaxation of a homogeneous vertical distribution of the Richardson number and turbulent viscosity coefficient and can be reduced to a single equation

$$\frac{\partial^2\nu_T}{\partial t^2} = \zeta\nu_0 N\left(2 - \frac{1}{\sigma_T}\right)\mathrm{Ri}_{\mathrm{cr}}\frac{\partial^2\nu_T}{\partial z^2}.$$

This equation implies that under the conditions formulated, large-scale perturbations of a uniform turbulence field will not spread in a diffusive way but propagate as running waves, moving upward or downward at a phase velocity

$$c_p = \sqrt{\zeta\nu_0 N\left(2 - \frac{1}{\sigma_T}\right)\mathrm{Ri}_{\mathrm{cr}}}$$

which is independent of the perturbation scale, that is, it carries no dispersion distortions. This is true in the case when the turbulent transfer of turbulence energy and its dissipation rate are negligibly small in comparison with their generation or dissipation due to changes in the Richardson number. By comparing the characteristic times of these processes for the periodic perturbation with the wave number K, namely $\tau_{\text{dis}} = 1/(\nu_0 K^2)$ for the former and $\tau_{\text{din}} = 1/(c_p K)$ for the second, we can obtain the condition of dominance of the wave propagation of perturbations over the diffusion process,

$$KL \ll 1. \tag{4.3.10}$$

This means that the wave propagation occurs for perturbations with a characteristic scale that greatly exceeds the macroscopic turbulence scale. As we have mentioned earlier, this scale in the self-similar mode is proportional to $\sqrt{b}$, so that for small b we can expect condition (4.3.10) to be satisfied.

The wave relaxation of inhomogeneity may also take place in situations in which the non-perturbed turbulence distribution is maintained owing to its generation by an external driving force. This driving force compensates for turbulence decay when the uniformly distributed Richardson number is only slightly greater than Ri_{cr}: $\text{Ri} = \text{Ri}_{\text{cr}} + \delta\,\text{Ri}_0$, $0 < \delta\,\text{Ri}_0 \ll \text{Ri}_{\text{cr}}$. In fact, the range of wave numbers is then bounded not only by condition (4.3.10) from above but also by the condition $KL \gg |\delta\,\text{Ri}_0|$ from below. If this last condition is satisfied, the oscillation period is small in comparison with the time of relaxation of uniform turbulence to the non-perturbed level that is given by the balance of generation and decay. Adding the generation rate P_0 to equation (4.3.7) and characterizing the non-perturbed level of turbulence by that corresponding to the turbulent viscosity ν_0 (related by formula (4.3.8) to the level of turbulence energy), we obtain

$$\nu_0 = \frac{c_* P_0}{\zeta N^2 \delta\,\text{Ri}_0}. \tag{4.3.11}$$

Expression (4.3.11) makes it possible to evaluate the minimum value of P_0 required to sustain turbulence: it must be sufficiently high for the value of ν_0 to stay much higher than the molecular viscosity ν. Furthermore, the same value of P_0 may be sufficient to sustain turbulence in the layer of reduced stability where $\delta\,\text{Ri}_0$ is small but insufficient in other layers where $\delta\,\text{Ri}_0 / \text{Ri}_{\text{cr}} = O(1)$. We assume that the thickness h of the reduced stability layer, determining the characteristic scale of vertical inhomogeneity distribution of non-perturbed values of $\delta\,\text{Ri}_0$ and ν_0, is much greater than the macroscopic scale of turbulence, which corresponds to the minimum possible level of turbulence energy. Then the relaxation of perturbation inhomogeneity induced by the wave will evolve mostly at the expense of the change in the mean flow due to turbulent transfer; the turbulence self-diffusion plays only a secondary role. For this reason, the locality condition for turbulence evolution in the field of the wave can be obtained from equation (4.3.9). For the mean flow not to change appreciably over one period of the wave it is necessary to restrict the growth of turbulent viscosity in the destabilizing phase by the condition

$$\int_t^{t+T} \nu_T(t)\,\mathrm{d}t \ll \frac{h^2\delta\,\mathrm{Ri}_0}{(2-1/\sigma_T)\,\mathrm{Ri}_{\mathrm{cr}}}. \tag{4.3.12}$$

If condition (4.3.12) is met, the evolution of turbulence in the wave field with the spatial period much greater than the thickness of the reduced viscosity layer can be described, in the reference frame in which $U(z_0) = 0$, by the equation

$$\frac{\mathrm{d}b}{\mathrm{d}t} = -\zeta b N(\mathrm{Ri} - \mathrm{Ri}_{\mathrm{cr}}) + P_0 \tag{4.3.13}$$

in which the rate of generation of turbulent energy by an external force P_0 is assumed to be sufficient to sustain turbulence at the stabilizing phase of the wave; on the other hand, it is assumed to be sufficiently small so as not to affect turbulence evolution at a destabilizing phase. These bounds may be satisfied at the same time, provided $\mathrm{Ri}_{\mathrm{cr}} \gg \nu_T(2-1/\sigma_T)/h^2$; this condition is, in its turn, in agreement with the assumption on the smallness of $\delta\,\mathrm{Ri}_0$ if the thickness of the reduced stability layer and the wave period are related by a formula that ensures a sufficiently high value of the corresponding Reynolds number $\mathrm{Re}_{\mathrm{eff}} = h^2/(\nu T)$.

We can assume in this situation that wave field perturbations due to turbulent transfer are small and the wave-induced deviations in the Richardson number $\delta\,\mathrm{Ri}_w$ are determined by the spatial wave structure non-distorted by the effect of turbulence. We also assume that the wave amplitude is sufficiently small for its nonlinear self-action not to have enough time (a) for manifesting itself over the time of interaction with turbulence it induces, (b) for the non-perturbed wave structure to be dictated by the solution of the linear boundary problem and (c) for the effect of turbulence to be expressed as a small correction. In this case the internal wave can be described by the distribution of the wave-induced vertical displacements of the fluid:

$$\eta(x, z, t) = [A(t)F(z) + \eta_1(t, z)]\mathrm{e}^{\mathrm{i}(\omega t - kx)} + \mathrm{c.c.}$$

where $F(z)$ describes the vertical mode structure of the non-perturbed wave, $\eta_1(t, z)$ is a small correction due to turbulence, $A(t) \equiv |A|\mathrm{e}^{\mathrm{i}\varphi}$ is the complex wave amplitude, which can vary in response to turbulence with a characteristic time that is much longer than the wave period $T = 2\pi/\omega$, x is the horizontal coordinate corresponding to the wave component $U(z)$ whose sign was chosen from the condition $U'(z_0) > 0$, and c.c. stands for the complex-conjugate variable. In addition to the constraints on frequency and wave number, we will also assume that its local frequency $\Omega(z) = \omega - kU(z)$ does not change sign nor vanishes over the entire thickness of the flow.

The deviation of the Richardson number from the non-perturbed value in the reduced stability layer can be written for this type of wave as

$$\delta\,\mathrm{Ri}_W = -2|A|S(z)\cos(\omega t - kx + \varphi) \tag{4.3.14}$$

where

$$S(z) = 2\frac{U'(z)}{c - U(z)} \mathrm{Ri}_{\mathrm{cr}} F(z) - cF'(z).$$

If the amplitude A is constant, equation (4.3.13) for the turbulence energy has a periodic solution

$$b(t) = \frac{P_0}{1 - \exp(\zeta N \delta \,\mathrm{Ri}_0 T)} \int_{t-T}^{t} \exp\left(-\zeta N \int (\delta \,\mathrm{Ri}_W(t'') + \delta \,\mathrm{Ri}_0)\, \mathrm{d}t''\right) \mathrm{d}t' \tag{4.3.15}$$

that can be used to find all turbulence parameters including the turbulent viscosity ν_T, for the current value of the complex wave amplitude $A(t)$ which is slowly varying due to turbulence.

This effect can be found using the standard asymptotic procedure of the 'slow' perturbations method (for details, see Moiseev *et al* 1984). This procedure allows one to obtain an equation for the magnitude of the complex wave amplitude from the constraint η_1:

$$2J \frac{\mathrm{d}|A|}{\mathrm{d}t} = \int_{z_1}^{z_2} \mathrm{Re}\left\{\mathrm{e}^{-\mathrm{i}\varphi} \nu_T^{(\omega,k)}\right\} U'(z) G(z)\, \mathrm{d}z \tag{4.3.16}$$

where

$$J = 2\int_{z_1}^{z_2} \Omega \left[F^2 + \left(\frac{F'}{k}\right)^2\right] \mathrm{d}z$$

z_1 and z_2 are the lower and upper boundaries of the flow, respectively,

$$\nu_T^{(\omega,k)}(z) = \frac{\omega k}{(2\pi)^2} \int_t^{t+T} \mathrm{d}t' \int_x^{x+2\pi/k} \nu_T(x', z, t') \exp(\mathrm{i}(kx' - \omega t'))\, \mathrm{d}x' \tag{4.3.17}$$

is the turbulent viscosity component in resonance with the wave, and

$$G(z) = \frac{U'}{\Omega}\left[\mathrm{Ri}_{\mathrm{cr}}\left(1 + \frac{1}{\sigma_T}\right)\frac{U'F}{c - U} - \left(2 - \frac{\mathrm{Ri}_{\mathrm{cr}}}{\sigma_T}\right)F'\right].$$

Therefore, the condition of wave amplitude growth is the positivity of the integral in the right-hand side of equation (4.3.16). As follows from equations (4.3.14), (4.3.15) and (4.3.17), the sign of the quantity $\mathrm{Re}\left\{\exp(-\mathrm{i}\varphi)\nu_T^{(\omega,k)}\right\}$ is determined by the sign of the function $S(z)$. If the thickness of the reduced stability layer is small in comparison with the characteristic scale of the vertical wave strucure, the sign of the integral is determined by that of the integrand at $z = z_0$. To be specific, we assume that $\Omega > 0$ (the direction of wave propagation is then given

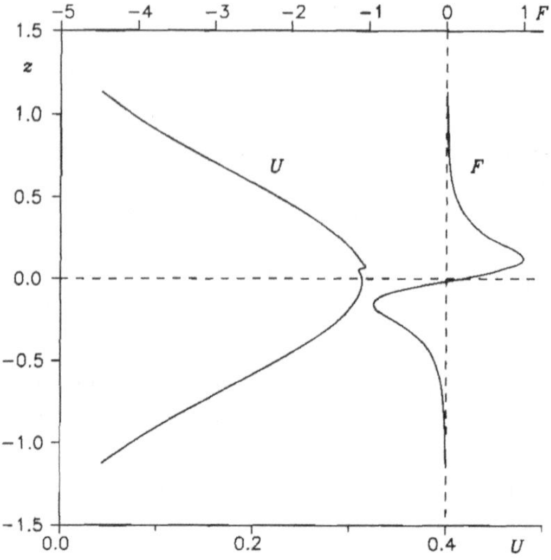

Figure 4.6. An example of vertical profiles of the normalized mean flow velocity $U(z)$ and the vertical wave displacement $F(z)$, satisfying the condition of WTI development for uniform vertical distribution of the Brunt–Väisälä frequency N, with arbitrary normalization scale of the coordinate H and normalization scale of the velocity $U_0 = NH$.

by the sign of the wave number k and phase velocity c); then the condition of wave amplification due to interaction with the wave-induced turbulence, that is, with the wave-turbulence instability, can be written as

$$\left(\frac{F'}{F} - 2\frac{U'}{c-U}\,\mathrm{Ri_{cr}}\right)\left[\mathrm{Ri_{cr}}\left(1+\frac{1}{\sigma_T}\right)\frac{U'}{c-U} - \left(2-\frac{\mathrm{Ri_{cr}}}{\sigma_T}\right)\frac{F'}{F}\right] > 0 \tag{4.3.18}$$

where the values of all functions are taken at $z = z_0$.

This condition is easily realizable if the function $F(z)$ has at least two extremums of opposite sign, thus passing through zero at some point in between the extremums. Then the ratio F'/F between the extremum points runs through all values from $-\infty$ to $+\infty$, and among them the values satisfying condition (4.3.18). If the reduced stability level lies at the level where F'/F runs through the range of values given by condition (4.3.18), the wave amplitude will grow. Figure 4.6 shows an example of a $U(z)$ profile which provides the necessary internal wave structure for WTI for vertically uniform distribution of N.

Comparing equations (4.3.9) and (4.3.16), we can establish that the reduced stability level is destroyed by the wave-induced turbulence much faster than the wave amplitude increases. However, if wave motions are concentrated in a wave packet that is spatially bounded along the x-axis, then this packet, moving along the reduced stability level, may reach a non-destroyed part of the layer and will be further enhanced. This process can be most efficient if Ri_0 increases in the direction of wave propagation, thus ensuring the most favorable relation between the wave growth rate and the rate of destruction of the reduced stability level.

The turbulence intensity should grow together with the wave amplitude because as a result of wave action, the parameters of the reduced stability level drift gradually farther beyond the turbulence generation threshold. Therefore, in this case the secondary instability not only involves new degrees of freedom in the relaxation process but even accelerates the relaxation in those degrees of freedom that were active from the start.

References

Albert J M, Simolon P L and Sudan R N 1991 *J. Geophys. Res.* **96** 16015–24

Barenblatt G I 1990 *J. Fluid Mech.* **212** 497–6

Buneman O 1963 *Phys. Rev. Lett.* **10** 285

Chimonas G 1972 *Bound.-Layer Met.* **2** 444–52

Farley D T 1963 *J. Geophys. Res.* **72** 895–901

Farley D T 1985 *J. Atmos. Terr. Phys.* **47** 729–44

Farley D T and Balsley B B 1973 *J. Geophys. Res.* **78** 227–39

Fua D, Chimonas G, Einaudi F and Zeman O 1982 *J. Atmos. Sci.* **39** 2450–63

Gibson M M and Launder B E 1976 *J. Heat Transfer* **198** 81–7

Gossard E E and Hook W H 1975 *Waves in the Atmosphere* (Amsterdam: Elsevier)

Hoh F C 1963 *Phys. Fluids* **6** 1184–91

Huba J D and Lee L C 1983 *Geophys. Res. Lett.* **10** 357–61

Kolmogorov A N 1942 *Izv. Akad. Nauk SSSR Ser. Fiz.* **6** 56–8 (in Russian)

Kolykhalov P I, Moiseev S S, and Pungin V G 1989 *Preprint No 1506* (Moscow: Space Research Institute) pp 1–19 (in Russian)

Lighthill M J 1952 *Proc. R. Soc.* **A211** 564–87

Maeda K, Tsuda T and Maeda H 1963 *Rep. Ionos. Space Res. Japan* **17** 3

Miles J W 1961 *J. Fluid Mech.* **10** 496–508

Moiseev S S and Pungin V G 1995 *Ukr. Phys. J.* **40** 427–9

Moiseev S S and Pungin V G 1996 *Zh. Exp. Teor. Fiz.* **110** 363–78

Moiseev S S, Sagdeev R Z, Tur A V and Yanovsky V V 1981 *Zh. Exp. Teor. Fiz.* **80** 597–607

Moiseev S S, Pungin V G, Sagdeev R Z, Suyazov N V and Etkin V S 1982 *Abstract 2nd Congress Soviet Oceanologists (Sebastopol, 1982)* pp 63–4 (in Russian)

Moiseev S S, Suyazov N V and Etkin V S 1984 *Preprint No 905* (Moscow: Space Research Institute) pp 1–18 (in Russian)

Pfaff R F, Kelley M C, Kudeki E, Fejer B G and Baker K D 1987 *J. Geophys. Res.* **92** 13578–96

Rogister A and D'Angelo N 1970 *J. Geophys. Res.* **75** 3879–87

Ronchi C, Sudan R N and Similon P L 1990 *J. Geophys. Res.* **95** 189–200

Rosenbluth N N and Longmire C 1957 *Ann. Phys.* **1** 120

Simon A 1963 *Phys. Fluids* **6** 382–8

Sudan R N, Akinrimisi J and Farley D T 1973 *J. Geophys. Res.* **78** 240–8

Turner J S 1973 *Buoyancy Effects in Fluids* (Cambridge: Cambridge University Press)
Whitham G B 1974 *Linear and Nonlinear Waves* Part 1 (New York: Wiley)

Chapter 5

The helical mechanism of large-scale structure generation in continuous media

5.1 Introduction

We have already emphasized the importance of that type of secondary unstable processes which leads to spontaneous symmetry breaking in a system. We gave a specific example in which the secondary instability 'transformed' the originally one-dimensional process to a two-dimensional one. Even less trivial cases are known, for example, a significant change in the topological properties of velocity fields (Moiseev *et al* 1988a) and magnetic moment fields (Merzliakov and Moiseev 1987) as a result of secondary instability. Thus the initial stage of the convective process in the Earth's atmosphere leads to the generation of cells of comparable vertical and horizontal sizes, with a simple topological structure; the helicity is zero for such types of field (i.e. $(\boldsymbol{V} \cdot \operatorname{rot} \boldsymbol{V}) = 0$). However, the field of turbulent pulsations caused by Earth's rotation (generated by convection itself, or arising due to other factors) becomes gyrotropic, that is, it has non-zero mean helicity. When this turbulence in the Earth's atmosphere is taken into account, the second stage of the convective process may result: generation of large-scale structures with the horizontal scale much larger than the vertical scale, with topologically non-trivial, linked lines of flow (i.e. $\langle \boldsymbol{V} \rangle \cdot \operatorname{rot}\langle \boldsymbol{V} \rangle \neq 0$, where $\langle \boldsymbol{V} \rangle$ is the velocity of the large-scale motion). What makes this option possible? Moiseev *et al* (1983) discovered a new type of instability: a small-scale turbulent velocity field $\boldsymbol{V}'$, which on the average possesses helicity $(\langle(\boldsymbol{V}' \cdot \operatorname{rot} \boldsymbol{V}')\rangle \neq 0)$, generates a large-scale average $\langle \boldsymbol{V} \rangle$ that possesses a similar property. The importance of this result follows first of all from the fact that it has become the cornerstone of a new field of theoretical studies: the dissipative mechanism of generation of large-scale structures in continuous media in which the Hamiltonian, correlators etc contain 'helical' terms (see, for example, Moiseev *et al* 1988a, Merzliakov and Moiseev 1987, Frisch *et al* 1987). It is also essential since a number of natural phenomena and physical features stem from the property described above (for example, see Merzliakov and Moiseev 1987, Moiseev *et al* 1983b).

The generation of helical vortical structures was studied in maximum detail in non-conducting media; this unavoidably affected the choice of material for this chapter.

We need to emphasize that one of the primary tasks in expanding this field is to carry out a similar analysis for the plasmas, where helicity is an inherent concept. The plasmas possessing this property are, for example, the plasma in non-uniform magnetic field, rotating plasma, etc. It is to be expected that 'helical' terms will arise in the correlators of turbulent magnetized plasmas (for example, when taking into account the helicone (whistler) oscillation branch). It seems to be important, therefore, to build for plasma-physics experts a picture of the state-of-the-art and of the promising directions of development in this very active field. We believe that this chapter will be of special importance for people working in geophysical hydrodynamics (of ionosphere, atmosphere, oceans and liquid core of the Earth). Suffice it to note in this connection that a catastrophe in nature is typically a large-scale process which describes a transition to a qualitatively new state; very often, the scenario follows the lines of the evolution of a secondary instability, namely, an instability 'prepares' the conditions for a large-scale secondary instability. Remarkably, owing to Earth's rotation, the helical mechanism is one of the dominant factors for the self-organization of the natural environment in the process of transforming from stage one to stage two.

This chapter is organized as follows. Section 2 treats a qualitative analysis of the fields of applicability of the large-scale structure generation approach using the helical mechanism. Sections 3 to 5 study this phenomenon in greater detail, applying the treatment to compressible and rotating media and also to shear flows. Finally, section 6 is devoted to a qualitative and experimental analysis of applicability of the helical mechanism to triggering of tropical cyclones and to modulation of the rotation frequency of the Earth.

5.2 Helical structures and helical mechanisms of generation in hydrodynamics

5.2.1 On the role of helicity, on the similarity and differences between the magnetic and vortex dynamo in turbulent media

A turbulent ensemble often possesses the gyrotropic property (helicity); for example, a random velocity field possesses this property if the medium is stratified and rotating, as happens with the Earth's atmosphere. Indeed, let a convective cell in the atmosphere float upwards. As the cell rises, it reaches cooler layers, expands and is forced to rotate by the Coriolis force. It is not difficult to see that $G = (V \operatorname{rot} V) \neq 0$. In a sinking cell, the sequence is reversed and for this reason the sign of G remains unchanged. We can conclude that the turbulent convective ensemble possesses a non-zero average helicity $\langle G\rangle = \langle(\boldsymbol{V}\,\mathrm{rot}\,\boldsymbol{V})\rangle$. In this case the spatial part of the correlation function for velocities $K_{\alpha\beta}(\boldsymbol{r},\boldsymbol{r}') = V_\alpha(\boldsymbol{r})V_\beta(\boldsymbol{r}')$, assuming the velocity field to be isotropic and uniform, can be written in the following general form (see, for example, Moiseev *et al* 1983b):

$$K_{\alpha\beta}(\boldsymbol{r},\boldsymbol{r}') = A(|\boldsymbol{r}-\boldsymbol{r}'|)\delta_{\alpha\beta} + B(|\boldsymbol{r}-\boldsymbol{r}'|)x_\alpha x_\beta + C(|\boldsymbol{r}-\boldsymbol{r}'|)\varepsilon_{\alpha\beta\gamma}x_\gamma \qquad (x_\alpha = r_\alpha - r'_\alpha). \tag{5.2.1}$$

Here $\delta_{\alpha\beta}$ is the Kronecker symbol and $\varepsilon_{\alpha\beta\gamma}$ is the antisymmetric tensor. Multiplying (5.2.1) by $\varepsilon_{\alpha\beta\gamma}\partial/\partial x_\delta$ and performing convolution over the indices α and β, we obtain $C(0) = -(1/3)G$. A random velocity field of this type is unstable and, as shown previously by Steenbeck *et al* (1966), generates in a conducting MHD medium a large-scale magnetic field $\langle\boldsymbol{H}\rangle$. It was shown much later (Moiseev *et al* 1983a) that a similar velocity field can generate in a non-conducting medium a large-scale helical vortex. This time gap of almost twenty years is not accidental, especially if we take into account that Krause and Rudiger (1974) proved a theorem which states that it is forbidden for a small-scale helical velocity field to generate a large-scale helical velocity field in an incompressible non-stratified fluid. In order to clarify the difference between the generation of $\langle\boldsymbol{H}\rangle$ and $\langle\boldsymbol{V}\rangle$, we will follow Artsimovich and Sagdeev (1979) and discuss at a qualitative level the peculiarities of generation of magnetic fields. If a turbulent small-scale flow generates a significantly smoother magnetic field (of a scale $L \gg l$), the problem gets considerably simpler. Let us write the magnetic field as

$$\boldsymbol{H}(\boldsymbol{r},t) = \langle\boldsymbol{H}(\boldsymbol{r},t)\rangle + \boldsymbol{h}'(\boldsymbol{r},t) \tag{5.2.2}$$

where $\langle\boldsymbol{H}(\boldsymbol{r},t)\rangle$ is the part of the magnetic field averaged over the small-scale ensemble and $\boldsymbol{h}'$ is a rapidly varying small-scale response to the motion of the medium. Using the familiar equation for the magnetic field in the MHD approximation, we write $\boldsymbol{h}'$ in the form

$$\frac{\partial\boldsymbol{h}'}{\partial t} = \mathrm{rot}[\boldsymbol{V}'\times\langle\boldsymbol{H}\rangle] + \eta_\mu\Delta\boldsymbol{h}' \tag{5.2.3}$$

where $\eta_\mu = c^2/(4\pi\sigma)$ is the magnetic viscosity (c is the speed of light and σ is the conductivity). Recalling that $\langle\boldsymbol{H}\rangle$ varies slower than $\boldsymbol{V}^t$ and assuming for simplicity that magnetic viscosity is sufficiently high, we can rewrite (5.2.3) as

$$(\langle\boldsymbol{H}\rangle\boldsymbol{\nabla})\boldsymbol{V}^t = \eta_\mu\langle k^2\rangle\boldsymbol{h}' \tag{5.2.4}$$

where $\langle k^2\rangle$ is a wave vector squared, averaged over the spectrum of turbulent pulsations.

The equation for the slowly varying component of the magnetic field can also be derived from the magneto-hydrodynamic equation for the magnetic field, after we average it over small-scale pulsations:

$$\partial\langle \boldsymbol{H}\rangle/\partial t = \mathrm{rot}\langle \boldsymbol{V}^{t} \times \boldsymbol{h}]\rangle + \eta_{\mu}\boldsymbol{\Delta}\langle \boldsymbol{H}\rangle. \tag{5.2.5}$$

It will be useful later in this chapter to point to the antisymmetric structure of the first term in the right-hand side of (5.2.5), namely

$$\left\{\mathrm{rot}\langle[\boldsymbol{V}^{t} \times \boldsymbol{h}']\rangle\right\}_i = \partial/\partial x_k\langle[h'_k V^t_i - V^t_k h'_i]\rangle. \tag{5.2.5a}$$

Using (5.2.4) and (5.2.1), we can transform (5.2.5) to

$$\partial\langle \boldsymbol{H}\rangle/\partial t = \alpha\,\mathrm{rot}\langle \boldsymbol{H}\rangle + \eta_{\mu}\boldsymbol{\Delta}\langle \boldsymbol{H}\rangle \tag{5.2.6}$$

where $\alpha = -\langle \boldsymbol{V}\,\mathrm{rot}\,\boldsymbol{V}\rangle/\eta_{\mu}\langle k\rangle^2$. We seek the solution of (5.2.6) in the form $\langle \boldsymbol{H}\rangle = H_0\exp(\gamma t)(\cos qz, \sin qz, 0)$, which leads us to the following dispersion equation (MHD α-effect):

$$\gamma = |\alpha| q - \eta_{\mu} q^2. \tag{5.2.7}$$

We see that if q is sufficiently small, equation (5.2.6) has unstable solutions. Therefore, gyrotropic turbulence with $\langle(\boldsymbol{V}^{t}\,\mathrm{rot}\,\boldsymbol{V}^{t})\rangle \neq 0$ results in feedback among various components of the large-scale magnetic field; it can provide amplification and sustain the magnetic field. The generation term α rot is a function of the derivative of velocity and in this sense is analogous to the viscous term. However, since it is proportional to the first derivative of velocity, a perturbation amplified by this term is always possible. The analogy to viscosity is especially instructive if we take into account the helicity fluctuations, which are strong at the stability threshold. Let the helicity fluctuate with characteristic time τ_1, such that $\tau_2 \gg \tau_1 \gg \tau$, where τ_2 and τ are the characteristic times of generation of a large-scale structure and of correlation for the helicity component of the turbulent velocity field, respectively. In the approximation chosen above we have

$$\alpha = \alpha_0 + \alpha'(t) \qquad \langle\alpha'\rangle = 0 \qquad \langle\alpha'(t)\alpha'(t')\rangle \approx 2D\delta(t - t').$$

For the sake of simplicity we also assume the fluctuations to be Gaussian, which simplifies the averaging procedure (see, for example, Klyatskin 1980). By averaging equation (5.2.6) we easily obtain (Kraichnan 1976)

$$\partial/\partial t\langle\langle \boldsymbol{H}\rangle\rangle = \alpha\,\mathrm{rot}\langle\langle \boldsymbol{H}\rangle\rangle + (\eta - D)\boldsymbol{\Delta}\langle\langle \boldsymbol{H}\rangle\rangle. \tag{5.2.8}$$

The form of equation (5.2.8) not only confirms the above statement on the similarity of the viscosity and helicity effects but also points to a lowering of the stability threshold when taking into account helicity fluctuations (since the viscosity coefficient decreases).

The small-scale gyrotropic velocity field creates a mechanism of feedback between various components of the large-scale velocity field (to be precise, of its vortical component). In this sense, there is a complete analogy between $\langle \boldsymbol{H} \rangle$ and rot$\langle \boldsymbol{V} \rangle$. However, differences between the structures of the terms may lead to generation processes: in the case of magnetic field they can be recast to an antisymmetric form (see (5.2.5*a*) while in the case of the velocity field in incompressible non-stratified medium the relevant terms can be made symmetric. This explains the above-mentioned banning theorem (Krause and Rudiger 1974) which follows immediately from the structure of the hydrodynamical term. To make it clear in a simplified manner, we write the velocity $\boldsymbol{V}$ in the form

$$\boldsymbol{V} = \langle \boldsymbol{V} \rangle + \boldsymbol{V}^t + \boldsymbol{V}' \qquad (5.2.9)$$

where $\langle \boldsymbol{V} \rangle$ is the large-scale velocity field, $\boldsymbol{V}^t$ is the originally fixed small-scale part of the velocity field with non-zero mean helicity and $\boldsymbol{V}$ is the inhomogeneous supplement to $\boldsymbol{V}^t$ resulting from the interaction with $\langle \boldsymbol{V} \rangle$. Therefore, we have chosen for simplicity the two-scale expansion. The nonlinear term, averaged over small-scale turbulent pulsations is given in the incompressible case by

$$\langle V_k \, \partial V_j / \partial x_k \rangle = \partial / \partial x_k (\langle V_j^t V_k' + V_j' V_k^t \rangle). \qquad (5.2.10)$$

Now following, for example, Krause and Rudiger (1974), we have

$$\langle V_j^t V_k' \rangle = T_{jk}^{(0)} + T_{jkl}^{(1)} \langle V_l \rangle + T_{jklm} \nabla_l \langle V_m \rangle + \dots . \qquad (5.2.11)$$

In the case of uniform isotropic turbulence, expression (5.2.11) can depend, in addition to $\langle \boldsymbol{V} \rangle$, only on the absolute antisymmetrical tensor ε_{ikl} and the symmetrical tensor δ_{ij}, so that $T_{ij}^{(0)} \sim \delta_{jk}$, $T_{jkl}^{(2)} \sim \varepsilon_{jkl}$, $T_{jklm} \sim \delta_{kl}\delta_{im}$. As a result we find for (5.2.10)

$$\partial / \partial x_k (\langle V_j^t V_k + V_j V_k^t \rangle) = \nu_T \Delta \langle V_j \rangle. \qquad (5.2.12)$$

Here antisymmetrical terms $\propto \varepsilon_{jkl}$ cancel out in the symmetrical combination; ν_T is the turbulent viscosity. Therefore, in the present case the helicity does not affect the large-scale mean characteristic. In order for the helicity to enter the averaged equations of motion, it is necessary to drop either incompressibility or isotropy (or homogeneity). Thus the nonlinear term will not symmetrize if we deal with inhomogeneous flow. The symmetry is definitely broken in the case of convection as well (a favored direction exists), so that the generation of large-scale mean velocity is possible. It is also necessary to note that we discuss the mean velocity (or mean vortex) but certainly not their squares. The square quantities can be generated by turbulence, generally, without the helicity.

5.2.2 Generation of large-scale helical vortices in convective media

We have mentioned in the preceding section that the conditions of applicability of the banning theorem are not met for convection, and the generation of helical structures is possible. This case is also interesting for practical reasons since it allows one to decipher the mechanism of generation of typhoons (hurricanes)—large-scale helical vortices in atmosphere, with linked streamlines of flow. On the other hand, this example is of interest for plasma physics since its analysis leads to possible plasma generalizations.

We follow here the work of Moiseev *et al* (1988a, 1988b) and Sagdeev *et al* (1987). Formally, our analysis is qualitatively close to that for the magnetic field (see the previous section). However, both the set of equations and the boundary conditions are obviously more complicated and the derivation is consequently more cumbersome. In view of this, we can dwell here only on the formulation of the problem, several intermediate calculations and an analysis of the final set of equations.

We know well (see, for example, Gershuni and Zhukhovitskii 1972) that the simplest problem in the convection theory is that of a plane-parallel layer of incompressible fluid heated from below; the layer is described by the Navier–Stokes equation, heat balance equation, and the condition of fluid incompressibility. The set of equations is complemented with the equation of state in which we neglect the dependence of density ρ on pressure P: $\rho = \rho_0(1 - \beta T)$, where T is temperature, $\beta = -\rho_0^{-1}(\partial\rho/\partial T)$ is the thermal expansion coefficient. The stability analysis is carried out for perturbations of velocity $\boldsymbol{V}$, temperature θ and pressure p, against the background of the ground state $T_0(z)$, $P_0(z)$, where $\operatorname{grad} T_0(z) = -A\boldsymbol{e}$ (A is a positive constant; $\boldsymbol{e}$ is a unit vector pointing vertically upwards). The set of equations for perturbations in the Boussinesq approximation has the form (Belian *et al* 1990)

$$\partial V_i/\partial t - \nu\Delta V_i + V_k\nabla_k V_i + \nabla_i p/\rho - \beta\theta g e_i = 0 \tag{5.2.13}$$

$$\partial\theta/\partial t - \kappa\nabla\theta + V_k\nabla_k\theta - Ae_iV_i = 0 \tag{5.2.14}$$

$$\nabla_k V_k = 0 \tag{5.2.15}$$

where g is the gravitational acceleration, ν is the kinematic viscosity, and κ is the thermal conductivity. The simplest boundary conditions for the set (5.2.13)–(5.2.15) are the so-called joint boundary conditions (Γ denotes the upper or lower bounding surface of the layer):

$$\boldsymbol{V}\big|_\Gamma = 0 \qquad \partial V_x/\partial z\big|_\Gamma = \partial V_y/\partial z\big|_\Gamma = 0 \qquad \theta\big|_\Gamma = 0. \tag{5.2.16}$$

We shall introduce into the Navier–Stokes equation a random external force $\boldsymbol{F}$ which creates a small-scale helical turbulence, assuming $\langle\boldsymbol{F}\rangle = 0$. For simplicity, we assume this small-scale turbulence to be uniform, isotropic and stationary. In the coordinate representation, the correlator of this random velocity field in coordinates is known:

$$Q_{ij}^T(t_1 - t_2, \boldsymbol{k}) = B(t_1 - t_2, k)(\delta_{ij} - k_ik_j/k^2) + G(t_1 - t_2, k)\varepsilon_{ijl}k_l$$

$$\langle\boldsymbol{V}\operatorname{rot}\boldsymbol{V}\rangle \sim \int k^2 G(t_1 - t_2, k)\,\mathrm{d}k. \tag{5.2.17}$$

The main problem is to derive a closed averaged equation from the set (5.2.13)–(5.2.15). The simplest case is that of small Reynolds numbers Re: we can then explicitly calculate the Reynolds stress tensor. Even in this simplified formulation, we can identify the main effects produced. Assuming therefore the nonlinear terms in equations (5.2.13)–(5.2.15) to be small, that is, $\mathrm{Re} = u\lambda/\nu \ll 1$ (u is the characteristic velocity of turbulent pulsations and λ is the external turbulence scale), we can solve the equation for temperature perturbation by iterations and arrive at a single equation for velocity:

$$L_{ij}V_j = -D_\kappa P_{im}\nabla_k(V_k V_m) - \beta A g P_{im} e_m e_j \nabla_k(V_k D_\kappa^{-1} V_j) + F_i. \qquad (5.2.18)$$

Here L_{ij} is a linear operator encountered in the convection theory:

$$\begin{aligned} L_{ij} &= D_\nu D_\kappa \delta_{ij} - \beta A g P_{im} e_m e_j \\ D_\nu &= \partial/\partial t - \nu\Delta \qquad D_\kappa = \partial/\partial t - \kappa\Delta \end{aligned} \qquad (5.2.19)$$

where $P_{im} = \delta_{im} - \nabla_i\nabla_m/\Delta$ is the projection operator that excludes the potential component of the velocity field. The differential operators in the denominator are defined as the integral operators with the corresponding Green's functions. Using the representation (5.2.9), we can also rewrite $\langle \boldsymbol{V} \rangle$ as

$$\begin{aligned} \langle \boldsymbol{V} \rangle &= \langle \boldsymbol{V}_T \rangle + \langle \boldsymbol{V}_p \rangle \\ \langle \boldsymbol{V}_T \rangle &= \mathrm{rot}(\boldsymbol{e}\varphi) \\ \langle \boldsymbol{V}_p \rangle &= \mathrm{rot}\,\mathrm{rot}(\boldsymbol{e}\psi) \end{aligned} \qquad (5.2.20)$$

here, $\langle \boldsymbol{V}_T \rangle$ and $\langle \boldsymbol{V}_p \rangle$ are the toroidal and poloidal components, respectively, of the solenoidal velocity field $\langle \boldsymbol{V} \rangle$. It is also convenient to make the equations dimensionless by choosing the fluid layer thickness h for the length scale, $T = h^2/\nu$ for time, and finally, h/T for velocity. After a sufficiently cumbersome averaging procedure (for details, see Moiseev *et al* 1988a) we obtain for the scalar and pseudoscalar functions, respectively,

$$(\partial/\partial t - \Delta)\psi = -\mathrm{Ra}\, s\mu_1(\boldsymbol{e}\nabla)^2\varphi \qquad (5.2.21)$$

$$\begin{aligned} &(\partial/\partial t - \Delta)(\mathrm{Pr}\,\partial/\partial t - \Delta)\Delta\varphi - \mathrm{Ra}\,\Delta_\perp\varphi \\ &\quad = -\mathrm{Ra}\, s\{\mu_1(\mathrm{Pr}\,\partial/\partial t - \Delta)[\Delta_\perp - (\boldsymbol{e}\nabla)^2]\}\psi. \end{aligned} \qquad (5.2.22)$$

For simplicity, equations (5.2.21) and (5.2.22) are written in the approximations of small Prandtl number, and $\mathrm{Pr} = \nu/\kappa \ll 1$. Here $\mathrm{Ra} = \beta A g h^4/\nu\kappa$ is the Rayleigh number, $S = G_0/(30\pi \mathrm{Re}\lambda/h)$ is the helicity parameter of turbulence. Note that

$$G = G_0[u^2\lambda^4/(1+\lambda^2k^2)^2]\mathrm{e}^{-(t-s)/\tau} \qquad \mu_1 \sim \lambda^2/T\nu.$$

We seek the solution of equations (5.2.21) and (5.2.22) taking into account the boundary conditions and find, for example, for φ,

$$\varphi(z, \boldsymbol{r}_\perp, t) \sim \mathrm{e}^{\gamma t}\mathrm{e}^{\mathrm{i}\boldsymbol{k}_\perp \boldsymbol{r}_\perp}\sin \pi z. \qquad (5.2.23)$$

Substituting the solutions for φ and ψ into (5.2.21) and (5.2.22) and assuming $\gamma = 0$, we find the neutral stability curve:

$$\mathrm{Ra_{cr}}(k_\perp^2, S) = \frac{2\pi^4(1 + k_\perp^2/\pi^2)^3}{k_\perp^2/\pi^2 + \{k_\perp^2/\pi^2 + 4(\pi^3 S)^2(1 + k_\perp^2/\pi^2)^3\mu_1^2(1 - k_\perp^2\pi^2)\}^{1/2}} \tag{5.2.24}$$

It is clear that if the helicity parameter tends to zero, the neutral stability curve transforms into the familiar curve for ordinary convection (Gershuni and Zhukhovitskii 1972):

$$\mathrm{Ra_{cr}}(k_\perp^2, 0) = \pi^6(1 + k_\perp^2/\pi^2)^3/k_\perp^2. \tag{5.2.25}$$

This curve has a minimum at $\mathrm{Ra_{min}} = 27\pi^4/4$, which is achieved at $k_{\mathrm{min}}^2 = \pi^2/2$. As a result, familiar convective cells are generated, whose horizontal size is of the same order as the vertical size. An increase in the parameter S reduces $\mathrm{Ra_{min}}(S)$, and the position of the maximum then shifts towards smaller wave numbers $k_\perp$, that is, the horizontal size of the cells grows. When the helicity parameter reaches the value

$$S = S_0 = 1/4\pi^3\mu_1 \tag{5.2.26}$$

the position of the minimum on the stability curve $\mathrm{Ra_{cr}}(k_\perp^2, S)$ shifts to the point $k_{\perp\,\mathrm{min}} = 0$. This means formally that the horizontal size of instability is infinite. This signifies, in its turn, a complete restructuring of convection, and as a result the system gains by replacing numerous convective cells with a single large cell (vortex) whose size is in fact dictated by the horizontal inhomogeneity of the problem. It should be emphasized that as follows from the set (5.2.21) and (5.2.22), the vortex produced by the instability definitely contains, owing to the helicity of the turbulent field, linked toroidal and poloidal velocity fields, and this produces topologically non-trivial configurations of flow lines. Note also that calculations show that the time of creation of this type of helical structure is quite close, for the parameters of the real tropical terrestrial atmosphere, to the evolution time for hurricanes, and its characteristic size is of the order $\sqrt{Lh}$, where L is the radius of the heating area.

5.2.3 Helical mechanism of generation of large-scale structures in non-convective hydrodynamic media

The case discussed above of generation of large-scale structures in a turbulent convective medium pertains first of all to stratified continuous media, that is, media with temperature or density gradients. The question then arises: in what sense is the helical mechanism of interest as a cause of generation of large-scale structures in other natural or laboratory conditions? We will limit our discussion here to a brief analysis of some qualitative features of selected cases and refer the reader to the cited publications, and also to sections 5.3, 5.4 and 5.5 of this chapter.

5.2.3.1 *Interaction of a stationary flow with helical turbulence*

The presence of a flux $\boldsymbol{V}^0(z)$ cancels the ban on the generation of large-scale structures by helical turbulence in incompressible liquid. The linear stage of evolution of a large-scale field $\langle \boldsymbol{V} \rangle$ for small Reynolds numbers is described by the following equation (Gvaramadze *et al* 1987)

$$\frac{\partial \langle V_i \rangle}{\partial t} + \frac{\partial}{\partial x_k}(V_i^{(0)} \langle V_k \rangle + V_k^0 \langle V_i \rangle) + G_{ijk} \frac{\partial \langle V_j \rangle}{\partial x_k} = \nu \Delta \langle V_i \rangle - \frac{\partial}{\partial x_i} \left(\frac{\langle P \rangle}{\rho} \right) \tag{5.2.27}$$

where

$$G_{ijk} = \tfrac{1}{2} G_* \left(5 \frac{\partial V_i^0}{\partial x_l} \varepsilon_{lkj} + 5 \frac{\partial V_m^0}{\partial x_l} \varepsilon_{klm} \delta_{ij} - 3 \frac{\partial V_k^0}{\partial x_l} \varepsilon_{ilj} - 3 \frac{\partial V_l^0}{\partial x_j} \varepsilon_{ikl} \right)$$

(the parameter G_* is proportional to the mean helicity of small-scale turbulence), $\langle P \rangle$ is the mean pressure and ν_0 is the kinematic viscosity. Owing to the low level of turbulence, we have taken turbulence into account only via helicity (the last term on the left in (5.2.27)), which provides a dissipative feedback between various components of the large-scale mean value; in the long run, this causes the arising instability.

5.2.3.2 *Vortical instability of helical turbulence in two-component medium*

Belian *et al* (1990) considered generation of large-scale vortices by small-scale helical turbulence in an incompressible fluid filled with solid particles. Assuming that on the scales considered, solid particles form a continuous medium, we can use the equations of two-phase hydrodynamics (Nigmatulin 1987, Nemtsov and Eidman 1989). These equations and the subsequent manipulations will not be reproduced here. However, we need to emphasize one central, formal aspect, by writing the expression for the divergence of velocity $\boldsymbol{V}$ of the fluid:

$$\operatorname{div} \boldsymbol{V} = -(4/3\pi a^3) \operatorname{div} n(\boldsymbol{V}_s - \boldsymbol{V}) \tag{5.2.28}$$

where n is the concentration of solid particles of radius a, and $\boldsymbol{V}_s$ is the hydrodynamic velocity of the solid phase. Expression (5.2.28) describes the displacement of the fluid by the solid phase, which ensures non-zero divergence in the incompressible fluid, provided $\boldsymbol{V}_s \neq \boldsymbol{V}$. However, the generation of large-scale structures by a small-scale helical subsystem in a divergent medium is possible (Moiseev *et al* 1983a). Let us consider the case of $V_s \ll V$ and, assuming that correlation relations in a system of solid particles is negligible, derive the effect of generation of large-scale structure with the increment given by

$$\operatorname{Im} \omega \propto \alpha_1^2/(2\nu_0) \tag{5.2.29}$$

where $\alpha \propto na^3 \langle \boldsymbol{V} \operatorname{rot} \boldsymbol{V} \rangle$ (Belian *et al* 1990, for details see section 5.4). This instability may prove significant in the atmosphere over sandy deserts, in regions with a large number of various contaminants, etc.

5.2.3.3 Compressible media

We have already mentioned in the introduction to this chapter that Moiseev *et al* (1983a) discovered the hydrodynamic α-effect. This effect was demonstrated using compressible liquids as examples. Later Moiseev *et al* (1988c) and Druzhinin and Khomenko (1989) developed a more detailed analysis of the linear theory and reported first results on nonlinear saturation of the α-effect in compressible media. The equation for $\langle \operatorname{rot} \boldsymbol{V} \rangle$ coincides with (5.2.6).

5.2.3.4 Conclusion

The analysis given above demonstrated that virtually in all real types of hydrodynamic media, the generation of large-scale structures by small-scale turbulence is indeed possible. The factor of special importance is the helical nature of turbulence; thus, this is typical for the turbulence of atmospheres of rotating planets. It is now clear, at any rate, that we deal here with a fundamental mechanism which dictates numerous processes in the oceans and the atmosphere. Helicity as a property is characteristic not only of hydrodynamic media but equally of a number of other types of continuous media, and manifests itself both in random and in dynamic subsystems. For example, the first dynamic version of dissipative generation of coherent helicone structures was considered using magnets for case study (Merzliakov and Moiseev 1987). A similar instability can be expected for liquid crystals whose free energy is known to contain (see, for example, Molchanov *et al* 1985) a 'screw' term. The possible regions of influence of helicity in plasmas were already discussed in the preface.

As for hydrodynamic media, one of the more important directions of progress in the near future is the rapid vortical dynamo. Success achieved in similar studies of generation of magnetic field (see, for example, Molchanov *et al* 1985) allows us to anticipate the successful outcome of the research into generation of vortical structures at large Reynolds numbers, when the Lagrangian approach and frozen-in integrals are well applicable.

Obviously, the complete understanding of the role played by the helicity mechanism in problems of physics of continuous media will be achieved through experimental studies, especially in the natural environment. Such experiments carried out on the *Academician Korolev* research vessel give sufficient ground to believe that this role is very significant (Veselov *et al* 1989). We need to stress first of all that both in the case of a weak tropical depression (initial stage of formation of a large-scale vortical perturbation) and in a tropical storm there exists a reverse energy cascade: an energy flux from small to large scales. Note that in this case the helicity of the large-scale velocity field is greater by an order of magnitude than for a weak depression (see also sections 5.5 and 5.6).

The results obtained thus give us evidence that the concept of the helicity mechanism, discussed above, appears quite plausible.

5.3 Hydrodynamic α-effect in compressible media

5.3.1 Introduction

Moiseev *et al* (1983a) considered uniform isotropic helical turbulence in a compressible medium and were able to show that the averaged linearized equation describing vorticity has the same form as the equation describing the α-effect in the electrodynamics of continuous media, namely

$$\partial/\partial t\, \boldsymbol{\Omega} + \alpha \operatorname{rot} \boldsymbol{\Omega} = \nu \Delta \boldsymbol{\Omega} \qquad (5.3.1)$$

where $\boldsymbol{\Omega} = \operatorname{rot} \boldsymbol{V}$ and the constant coefficients α and ν are expressed in terms of the statistical parameters of the turbulent velocity field. The effect of generation of large-scale vortices is caused by the term $\alpha \operatorname{rot} \boldsymbol{\Omega}$ appearing in equation (5.3.1) and resulting in exponential growth of vorticity. The factor α in this term is expressed through turbulence helicity. Before this result was reported, Kraichnan (1973), Brissaud *et al* (1973) and Moffat (1981) argued that helicity may lead to energy transfer from small to large scales; however, these papers did not derive averaged equations and therefore, could not consider instability.

As follows from the papers mentioned in section 5.2, the α-effect is as much a natural property of helical turbulence in non-conducting fluid as the generation of magnetic fields by helical turbulence is in conducting fluid. The hydrodynamic α-effect ensures energy transfer from relatively small-scale vortices in helical turbulence to large-scale vortical structures, which determines the dynamics in many systems. We can thus say that the helical turbulence engages the mechanism of large-scale vortical dynamo. However, in contrast to the case of magnetic hydrodynamics, helicity is not sufficient for this mechanism to function: additional factors of symmetry breaking are required, resulting in many cases in a tensor nature of the α-effect. In the case of the compressible fluid, however, the coefficient α treated in this section remains scalar, and the equation of the α-effect takes the simplest form. Moiseev *et al* (1983a) derived the α-effect with a number of assumptions that considerably simplify the derivation of the averaged equations. One of these assumptions was the delta-correlated nature of the vortical part of turbulence. As a result the coefficient α in equation (5.3.1) became independent of the medium compressibility parameter. The question thus arises of the status of equation (5.3.1) with respect to compressible fluids and to the applicability of the vortical dynamo equation, because the α-effect must vanish in incompressible fluids (Krause and Rudiger 1974). In this section, following Moiseev *et al* (1988c), we derive the α-effect equations for compressible fluids without assuming the δ-correlated nature of turbulence and can study how the generation parameter α depends on the compression parameter of the medium. It will be shown below that this dependence is characterized by the parameter $\mu = \lambda_{\text{cor}}/c\tau_{\text{cor}}$, where c is the velocity of sound, and λ_{cor} and τ_{cor} are energy-carrying spatial and temporal scales of turbulence. A number of limiting cases are investigated; thus the coefficient $\alpha \sim \mu^2$ in the case of weak compressibility ($\mu \ll 1$) and vanishes in the limiting case of incompressible fluid $c \to \infty$. Large values of the parameter $\mu \gg 1$ correspond to short correlation time, and $\mu \to \infty$ in the case of processes that are δ-correlated in time. In this limiting case the generation coefficient α is independent of μ, that is, is independent of medium compressibility, which agrees with the results reported in Moiseev *et al* (1983a).

The structure of this section is the following. Subsection 5.3.2 presents qualitative arguments and subsection 5.3.3 outlines the method of deriving the averaged equation; in subsection 5.3.4 we calculate the coefficients of the equation for the model correlation function of turbulence, analyze their dependence on the compressibility parameter and the limiting cases of incompressible fluid and δ-correlated process. In subsection 5.3.5 we discuss large-scale instability and in subsection 5.3.6, the physical mechanism and the possible applications of the α-effect.

5.3.2 Qualitative discussion

We begin with the equations of motion of the fluid; if the compressibility of the fluid is taken into account, these equations become

$$\frac{\partial V_i}{\partial t} + V_k \frac{\partial V_i}{\partial x_k} = \nu_0 \Delta V_i - \frac{c^2}{\rho_0}\frac{\partial \rho}{\partial x_i} \tag{5.3.2}$$

$$\frac{\partial \rho}{\partial t} + \frac{\partial (V_k \rho)}{\partial x_k} = 0 \tag{5.3.3}$$

where ν_0 is the kinematic viscosity coefficient, c is the velocity of sound and ρ is density. To simplify things, we consider a polytropic gas with pressure $P = \rho^\gamma$ and polytrope exponent $\gamma = 2$. Assume now that owing to external perturbations (or in a different manner) turbulence is produced in a medium described by equations (5.3.2) and (5.3.3). We denote the velocity of turbulent pulsations by $\boldsymbol{V}_1^t$. Then the velocity and density can be shown as sums of regular and random components:

$$\begin{aligned}
&\boldsymbol{V} = \boldsymbol{V}^{(1)}(\boldsymbol{r}, t) + \boldsymbol{V}_1^t(\boldsymbol{r}, t)\\
&\langle \boldsymbol{V} \rangle = \boldsymbol{V}^{(1)}\\
&\langle \boldsymbol{V}_1^t \rangle = 0\\
&\rho = \rho_0 + \rho^{(1)}(\boldsymbol{r}, t) + \rho_1^t(\boldsymbol{r}, t)\\
&\langle \rho \rangle = \rho_0 + \rho^{(1)}(\boldsymbol{r}, t)\\
&\langle \rho_1^t \rangle = 0
\end{aligned}$$

where ρ_0 = const is the constant density that corresponds to the rest state $V = 0$, $\rho^{(1)}$ and $\rho_1^{(t)}$ are the regular and random components of the variable part of density. By averaging equations (5.3.2) and (5.3.3), we obtain

$$\frac{\partial V_i^{(1)}}{\partial t} + \left\langle V_{1k}^t \frac{\partial V_{1i}^t}{\partial x_k} \right\rangle + V_k^{(1)} \frac{\partial V_i^{(1)}}{\partial x_k} = \nu_0 \Delta V_i^{(1)} - \frac{c^2}{\rho_0} \frac{\partial \rho^{(1)}}{\partial x_i} \tag{5.3.4}$$

$$\frac{\partial \rho^{(1)}}{\partial t} + \frac{\partial (V_k^{(1)}(\rho_0 + \rho^{(1)}) + \langle V_{1k}^t \rho_1^t \rangle)}{\partial x_k} = 0. \tag{5.3.5}$$

These equations are not closed since they include unknown terms $\langle V_{1k}^t (\partial V_{1i}^t / \partial x_k) \rangle$ and $\langle V_{1k}^t \rho_1^t \rangle$ that describe the Reynolds stress. Our task is to express these terms via the parameters of the mean fields $\boldsymbol{V}^{(1)}$ and $\rho^{(1)}$ and the statistical parameters of the turbulence fields of velocity and density for the case of weak nonlinearity. We will follow this procedure explicitly in the next subsection using the functional techniques, and will only give qualitative arguments here.

First of all, we need to be more specific in the formulation of the problem.

(i) Let us be more precise about averaging, denoted by angle brackets. An average (mean) quantity is an average over an ensemble. For an interpretation of this averaging we can assume that the ergodic hypothesis holds.

(ii) We need to formulate the problem of evolution of large-scale solenoidal component of the field $\boldsymbol{V}^{(1)}$ in a turbulent medium; we assume that the characteristic temporal and spatial scales T and L of the field $\boldsymbol{V}^{(1)}$ are much greater than the energy-carrying turbulence scales τ and λ ($T \gg \tau$, $L \gg \lambda$). In general, the packet $\boldsymbol{V}^{(1)}$ may contain harmonics of various scales; furthermore, the initially large-scale packet $\boldsymbol{V}^{(1)}$ may become fuzzy as a result of interaction with turbulence but we neglect this phenomenon in this approximation, as well as the reverse effect of the field $\boldsymbol{V}^{(1)}$ on that part of turbulence that is produced by the external force. This means that in the approximation chosen here (we refer to it as two-scale approximation) the statistical characteristics of external turbulence are regarded as unchangeable and are treated as fixed, which appears to be justifiable at small amplitudes $\boldsymbol{V}^{(1)}$.

(iii) Since we assume the external turbulence to be fixed, we also assume it to be uniform and isotropic. This signifies that all its moments are invariant with respect to translation and rotation of the coordinate system, that is, there are no favored directions and positions in the medium.

In the problem formulated in this way, it is obvious that the quantity $\langle V_{1k}^t (\partial V_{1i}^t / \partial x_k) \rangle$ (the Reynolds stress) is a functional of $\boldsymbol{V}^{(1)}$ and, in general, that this dependence may be complicated and nonlinear. At small amplitudes $\boldsymbol{V}^{(1)}$, however, it can be linearized and then we approximately consider the quantity $\langle V'_{1k}(\partial V'_{1i}/\partial x_k)\rangle$ as a linear functional of $\boldsymbol{V}^{(1)}$. Therefore, the one-point average

$$\left\langle V'_{1k}(\boldsymbol{r},t)\frac{\partial V'_{1i}}{\partial x_k}(\boldsymbol{r},t)\right\rangle = T_i^{(0)} + T_{ik}^{(1)}V_k^{(1)}(\boldsymbol{r},t) + T_{ikl}\frac{\partial V_l^{(1)}(\boldsymbol{r},t)}{\partial x_k}$$
$$+ T_{ikl}^{(3)}\frac{\partial}{\partial x_k}\frac{\partial}{\partial x_l}V_m^{(1)}(\boldsymbol{r},t) + T_{iklmn}\frac{\partial}{\partial x_k}\frac{\partial}{\partial x_l}\frac{\partial V_n^{(1)}(\boldsymbol{r},t)}{\partial x_m}\dots \tag{5.3.6}$$

can be rewritten as a series expansion in gradients of the field $\boldsymbol{V}^{(1)}$, where the expansion coefficients (the tensors $\boldsymbol{T}^{(i)}$) must be written in terms of the moments of the external turbulent fields. Since we assume the medium to be uniform, the tensors $\boldsymbol{T}^{(i)}$ are constant and independent of coordinates. As the medium is isotropic, these tensors must be invariant with respect to rotations, that is, must be constructed of invariant tensors δ_{ik} and ε_{ijk} (where δ_{ik} is the Kronecker symbol and ε_{ijk} is the absolutely asymmetric tensor) and of the scalar characteristics of turbulent fields. Therefore, by selecting only the solenoidal part of the Reynolds stress, required below, we obtain

$$\left\langle V_{1k}^{(t)}(\boldsymbol{r},t)\frac{\partial V'_{1i}(\boldsymbol{r},t)}{\partial x_k}\right\rangle = c_0V_i^{(1)} + c_1\varepsilon_{ijk}\frac{\partial V_k^{(1)}}{\partial x_i} + c_2\Delta V_i^{(1)}$$
$$+ c_3\Delta\varepsilon_{ijk}\frac{\partial V_k^{(1)}}{\partial x_j} + c_4\Delta^2V_i^{(1)} + \dots$$
$$= \varepsilon_{ijk}\frac{\partial}{\partial x_j}(c_1 + c_3\Delta + c_5\Delta^2 + \dots c_{2n+1}\Delta^n + \dots)V_k^{(1)}$$
$$+ (c_0 + c_2\Delta + c_4\Delta^2 + \dots + c_{2n}\Delta^n + \dots)V_i^{(1)} \tag{5.3.7}$$

where the quantities $c_0, c_2, c_4 \dots, c_{2n}$ are scalar constants and $c_1, c_3, c_5 \dots, c_{2n+1}$ are pseudoscalar constants that describe a fixed turbulent field. Obviously, the quantity $\langle V_{1k}^{(t)}(\partial V'_{1i}/\partial x_k)\rangle$ is a polar vector; for the quantity $c_1\varepsilon_{ijk}(\partial V_k^{(1)}/\partial x_j)$ to be a polar vector as well, not an axial vector, it is necessary that the constant c_1 be a pseudoscalar. The constant c_0 must vanish, otherwise the uniform turbulence would contribute to the Reynolds stress.

We will now evaluate the coefficients c_i in the correlation approximation, that is, we assume that the properties of turbulence are dictated completely by the second-order correlation function.

The correlation tensor of non-perturbed turbulent pulsations of uniform isotropic turbulence can be written in explicit form (see, for example, Monin and Yaglom 1965):

$$\langle V'_{1i}(\boldsymbol{r}_1,t)V'_{1j}(\boldsymbol{r}_2,t)\rangle = A(r,\tau)\delta_{ij} + B(r,\tau)r_ir_j + H(r,\tau)\varepsilon_{ijk}r_k \tag{5.3.8}$$
$$\boldsymbol{r} = \boldsymbol{r}_1 - \boldsymbol{r}_2 \qquad \tau = t_1 - t_2.$$

We have shown the correlator of only the vortical part of turbulence,[1] assuming that the condition $(\partial\langle V^t_{1i}V^t_{1j}\rangle)/\partial x_i = 0$ is satisfied; as a result the quantity $B(\boldsymbol{r},\tau)$ is not independent and is related to $A(\boldsymbol{r},\tau)$. There is only one pseudoscalar quantity in the correlator, namely $A(\boldsymbol{r},\tau)$ (helicity $\langle H(0,0)\rangle = (1/6)\langle \boldsymbol{V}^t_1(\boldsymbol{r},t)\,\mathrm{rot}\,\boldsymbol{V}^t_1(\boldsymbol{r},t)\rangle$). Then the dimension-based arguments give us

$$c_1 \sim \int H(0,\tau)\,\mathrm{d}\tau \sim H(0,0)\tau_{\mathrm{cor}} \tag{5.3.9}$$

$$c_2 \sim \int A(0,\tau)\,\mathrm{d}\tau \sim A(0,0)\tau_{\mathrm{cor}}. \tag{5.3.10}$$

One important factor must be mentioned. All lower-order terms of expansion (5.3.7) beginning with c_3 will be smaller than the preceding ones; this follows from dimensional arguments. Indeed, $c_3 \sim \lambda^2 c_1$, $\Delta\boldsymbol{V}^{(1)} \sim L^{-2}\boldsymbol{V}^{(1)}$ and therefore,

$$c_3\Delta\frac{\partial V_k^{(1)}}{\partial x_j} \sim (\lambda/L)^2 c_1\frac{\partial V_k^{(1)}}{\partial x_j}$$

$$c_4\Delta^2 V_i^{(1)} \sim (\lambda/L)^2\Delta V_i^{(1)}$$

$$\vdots$$

This means that the series (5.3.6) is an expansion in a small parameter λ/L and can be truncated, retaining only the first terms.

The form of expansion (5.3.7) implies that the term proportional to the constant c_1 is in fact the term of novel nature in the equation for vorticity that leads to the α-effect. The term proportional to c_2 gives a dissipation term in the equation for vorticity, and the estimate (5.3.10) gives the turbulent viscosity. Expansion (5.3.7) permits another important conclusion: for the α-effect to exist, turbulence must be mirror-noninvariant. No doubt, its moments remain true tensors; there exist, however, certain characteristics by measuring which we can distinguish between the right-handed and the left-handed coordinate systems (e.g. by measuring the correlation function of the scalar product of velocity and vorticity $\langle \boldsymbol{V}\,\mathrm{rot}\,\boldsymbol{V}\rangle$ in helical turbulence). A familiar example of mirror-noninvariant turbulence is helical turbulence. However, helicity alone (or another mirror-noninvariant characteristic alone) is not sufficient; additional symmetry breaking is required. It can be, for example, compressibility-caused breaking of symmetry of the tensor $T^{(2)}_{ikl}$, which holds in incompressible fluids (this can be seen from the expansions (5.3.6) and (5.3.7)). Indeed, the nonlinear

[1] It will be shown below that the effects of interest to us here follow from the last term in (5.3.8) which 'automatically' satisfies the compressibility condition. Therefore, we simply do not write the contribution to the correlator from the potential component of velocity $\boldsymbol{V}^t$ and thus avoid overburdening the manipulations. Taking this contribution into account is practically immaterial to the aspect we are discussing now.

term $\langle V^t_{1k}(\partial V^t_{1i}/\partial x_k)\rangle$ in the case of incompressible fluids (when $(\partial/\partial x_k)V_k = 0$) can be written as $\partial\langle V^t_{1k}V^t_{1i}\rangle/\partial x_k$. The tensor $\langle V^t_{1k}V_{1i}\rangle$ is symmetrical with respect to indices and we immediately find that the tensor $T^{(2)}_{ikl}$ must also be symmetrical with respect to the indices i, k and cannot equal ε_{ikl}. A tensor of rank three cannot be constructed out of δ_{ik} only, so that there is no hydrodynamic α-effect (referred to hereafter as simply α-effect) in isotropic turbulence of incompressible fluids (Krause and Rudiger 1974). The α-effect in incompressible fluids does realize, however, if one of the symmetries is violated, for example, by adding a non-uniform flux (Gvaramadze *et al* 1989, Tur *et al* 1987) or a temperature gradient in the gravitational field (Moiseev *et al* 1988a, Sagdeev *et al* 1987). In this case the tensors $\boldsymbol{T}^{(i)}$ need not be symmetrical with respect to rotations and translations and can be expressed in terms of not only the invariant tensors ε_{ikl} and δ_{ik} but also of the characteristics of inhomogeneity or anisotropy. It must nevertheless be emphasized that the results presented in this subsection are no more than 'navigational' hints; the closed equation will be much more rigorously derived in the next subsection, and we obtain an explicit expression for the coefficient α in the averaged equation for vorticity.

5.3.3 Closure of averaged equation

Assume that turbulence has been generated in a compressible medium by a random external force $\boldsymbol{F}'$ ($\langle \boldsymbol{F}'\rangle = 0$). The velocity of turbulent pulsations satisfies the equations

$$\frac{\partial V^t_i}{\partial t} + V^t_k\frac{\partial V^t_i}{\partial x_k} = \nu_0\Delta V^t_i - \frac{c^2}{\rho_0}\frac{\partial \rho^t}{\partial x_i} + F'_i \qquad (5.3.11)$$

$$\frac{\partial \rho^t}{\partial t} + \frac{\partial}{\partial x_k}(V^t_k\rho^t) = 0 \qquad (5.3.12)$$

where $\rho_0 = \text{const}$ is a constant velocity, and ρ^t is the fluctuating component of density, caused by the turbulent motions in the medium ($\langle\rho\rangle = \rho_0$, $\langle\rho^t\rangle = 0$). Assume that the properties of the random external force are such that turbulence is uniform and isotropic. The correlation tensor (or covariance matrix) of the field $\boldsymbol{V}^t$ is then of the form (5.3.8), and the correlation function $\langle\rho^t\boldsymbol{V}^t\rangle$ is zero. The uniform isotropic turbulence satisfying equations (5.3.11) and (5.3.12) is the ground state. Note that the correlation function (5.3.8) describes correlations of the solenoidal component of the velocity field $\boldsymbol{V}^t$.

Assume now that turbulence is perturbed by a large-scale vortical perturbation $\boldsymbol{V}^{(1)}$ which interacts with the turbulent fields and consequently these last acquire inhomogeneity additional terms $\boldsymbol{V}'$ and ρ'. The velocity and density can be written as

$$\begin{aligned}
&\boldsymbol{V} = \boldsymbol{V}^{(1)}(\boldsymbol{r}, t) + \boldsymbol{V}^{t}(\boldsymbol{r}, t) + \boldsymbol{V}'(\boldsymbol{r}, t)\\
&\langle \boldsymbol{V} \rangle = \boldsymbol{V}^{(1)}(\boldsymbol{r}, t)\\
&\langle \boldsymbol{V}^{t} \rangle = \langle \boldsymbol{V}' \rangle = 0\\
&\rho = \rho_0 + \rho^{(1)}(\boldsymbol{r}, t) + \rho^{t}(\boldsymbol{r}, t) + \rho'(\boldsymbol{r}, t)\\
&\langle \rho \rangle = \rho_0 + \rho^{(1)}(\boldsymbol{r}, t)\\
&\langle \rho^{t} \rangle = \langle \rho' \rangle = 0.
\end{aligned} \tag{5.3.13}$$

The equations describing the mean fields $\boldsymbol{V}^{(1)}(\boldsymbol{r}, t)$ and $\rho^{(1)}(\boldsymbol{r}, t)$ are written, taking into account equations (5.3.11) and (5.3.12), in the form

$$\frac{\partial V_i^{(1)}}{\partial t} + \left\langle V_k' \frac{\partial V_i^{t}}{\partial x_k} \right\rangle + \left\langle V_k^{t} \frac{\partial V_i'}{\partial x_k} \right\rangle = \nu_0 \Delta V_i^{(1)} - \frac{c^2}{\rho_0} \frac{\partial \rho^{(1)}}{\partial x_i} \tag{5.3.14}$$

$$\frac{\partial \rho^{(1)}}{\partial t} + \frac{\partial(\langle V_k^{t} \rho' \rangle + \langle V_k' \rho^{t} \rangle)}{\partial x_k} = 0. \tag{5.3.15}$$

We therefore assume that $\boldsymbol{V}^{t}$ and ρ^{t} are again described by equations (5.3.11) and (5.3.12) and that all changes in the random fields are described by the functions $\boldsymbol{V}'(\boldsymbol{r}, t)$ and $\rho'(\boldsymbol{r}, t)$, which satisfy equations

$$\frac{\partial V_i'}{\partial t} - \nu_0 \Delta V_i' + \frac{c^2}{\rho_0} \frac{\partial \rho'}{\partial x} = -V_k^{(1)} \frac{\partial V_i^{t}}{\partial x_k} - V_k^{t} \frac{\partial V_i^{(1)}}{\partial x_k} \tag{5.3.16}$$

$$\frac{\partial \rho'}{\partial t} + \rho_0 \frac{\partial V_k'}{\partial x_k} = -\frac{\partial}{\partial x_k}(V_k^{(1)} \rho' + V_k^{t} \rho^{(1)}). \tag{5.3.17}$$

In these equations we dropped the terms nonlinear in the perturbations $\boldsymbol{V}^{(1)}$, $\rho^{(1)}$ and $\boldsymbol{V}'$, ρ' which we assumed to be small (this is true for small amplitudes of the mean fields $\boldsymbol{V}^{(1)}$ and $\rho^{(1)}$). The averaged equations (5.3.14) and (5.3.15) are not closed because they contain unknown terms in angle brackets (the Reynolds stress). To close the averaged equations, we make use of the Furutsu–Novikov formula which relates the mean value of the product of the random process φ by the functional $F[\varphi]$ to the integral of the product of the correlation function by the mean value of the variational derivative:

$$\langle F[\varphi]\varphi \rangle = \int \langle \varphi\varphi \rangle \left\langle \frac{\delta F[\varphi]}{\delta \varphi} \right\rangle \mathrm{d}t'.$$

In our case $\boldsymbol{V}'$ is a functional of the random processes $\boldsymbol{V}^{t}$ and ρ^{t}. The correlation function of these processes is known, so we only need to calculate the mean value of the variational derivative $\langle \delta \boldsymbol{V}'/\delta \boldsymbol{V}^{t} \rangle$ and substitute it into the Furutsu–Novikov formula, which in our case takes the form

$$\langle V_i'(\boldsymbol{r}_1, t) V_j^{t}(\boldsymbol{r}, t) \rangle = \int \langle V_j^{t}(\boldsymbol{r}, t) V_m^{t}(\boldsymbol{r}', t') \rangle \left\langle \frac{\delta V_i'(\boldsymbol{r}_1, t)}{\delta V_m^{t}(\boldsymbol{r}', t')} \right\rangle \mathrm{d}\boldsymbol{r}' \, \mathrm{d}t'. \tag{5.3.18}$$

The second term, which could be expected to appear in (5.3.18) (i.e. $\int \langle \boldsymbol{V}'\rho'\rangle \langle \delta \boldsymbol{V}'/\delta\rho'\rangle \, \mathrm{d}\boldsymbol{r}' \, \mathrm{d}t'$), vanishes because the correlation $\langle \boldsymbol{V}'\rho'\rangle$ vanishes owing to the uniformity of the fields $\boldsymbol{V}'$ and ρ'.

The correlation function $\langle V_j' V_m'\rangle$ in formula (5.3.18) is known, so it remains to find the mean value of the variational derivative and then calculate the integral. Details of the calculations can be found in Moiseev *et al* (1988c); here we only trace the calculation algorithm. Equations (5.3.16) and (5.3.17) are linear in the variables $\boldsymbol{V}'$, ρ' and are thus solvable via the Fourier transform. We thus find the explicit form of functional $\boldsymbol{V}'[\boldsymbol{V}']$, so that further calculation of the variational derivative is straightforward.

Once the mean value $\langle V_i'(\boldsymbol{r}_1, t) V_j'(\boldsymbol{r}_2, t)\rangle$ has been calculated, we differentiate with respect to one of the variables and take the limit $\boldsymbol{r}_1 \to \boldsymbol{r}_2$. This gives us the one-point mean values $\langle V_k'(\boldsymbol{r}, t)(\partial V_i'(\boldsymbol{r}, t)/\partial x_k)\rangle$ and $\langle V_k'(\boldsymbol{r}, t)(\partial V_i'(\boldsymbol{r}, t)/\partial x_k)\rangle$. Applying the rot operation to equation (5.3.14), we arrive at a linear equation for vorticity, $\boldsymbol{\Omega} = \operatorname{rot} \boldsymbol{V}$:

$$\frac{\partial \boldsymbol{\Omega}}{\partial t} + \alpha \operatorname{rot} \boldsymbol{\Omega} = \nu \Delta \boldsymbol{\Omega} \tag{5.3.19}$$

$$\alpha = -\frac{8\pi}{3} \int_{-\infty}^{+\infty} \mathrm{d}\omega \int_0^{\infty} k^4 \, \mathrm{d}k \frac{G(k, \omega) \, \mathrm{i}\omega}{c^2k^2 - \omega^2 + \mathrm{i}\omega\nu_0 k^2} \tag{5.3.20}$$

where $G(k, \omega)$ is the Fourier transform of the helicity coefficient in the turbulence correlator (Moiseev *et al* 1988c). The turbulent viscosity coefficient is (Krause and Rudiger 1974)

$$\nu = \nu_0 + \frac{16\pi}{15} \int_{-\infty}^{+\infty} \mathrm{d}\omega \int_0^{\infty} \mathrm{d}k \, k^2 \frac{2\nu_0 k^2 - \mathrm{i}\omega}{(\nu_0 k^2 - \mathrm{i}\omega)^2} D(k, \omega) \tag{5.3.21}$$

where $D(k, \omega)$ is the scalar coefficient in the Fourier transform of the turbulence correlator.

Equation (5.3.19) is an implementation of the program outlined in section 5.3.2, and the quantities α and ν are the explicit expressions for the coefficients c_1 and c_2 of the expansion (5.3.7).

5.3.4 α as a function of compressibility coefficient: limiting cases

We will discuss in more detail the structure of the expression (5.3.20). The integrand function depends parametrically on the speed of sound and the kinematic viscosity coefficient ν_0. Furthermore, it is clear from dimension-based arguments that the helicity $G(k, \omega)$ also depends parametrically on the quantities r_{cor} and λ_{cor}. Then $G(k, \omega) = G(k\lambda_{\mathrm{cor}}, \omega\tau_{\mathrm{cor}})$, where λ_{cor} and τ_{cor} are the characteristic helicity parameters of the turbulent velocity field.

To further analyze the expression (5.3.20), we will need additional assumptions on the form of the spectral density of the correlation function of velocity and vorticity $G(k, \omega)$. To simplify the problem, we assume that the properties of the random external force are such that the function $G(k, \omega)$ is factorizable,

$$G(k, \omega) = g_0 G_1(\omega) G_2(k) \tag{5.3.22}$$

where g_0 is a dimensional constant, and G_1 and G_2 are dimensionless functions.

As an example, we consider a function $G_1(\omega)$, chosen as

$$G(\omega) = \left(\frac{\pi}{2}\right)^{1/2} \frac{\omega_{\rm cor}^2}{\omega^2 + \omega_{\rm cor}^2} \tag{5.3.23}$$

where $\omega_{\rm cor} \sim 1/\tau_{\rm cor}$. For $\tau_{\rm cor} \to 0$ and $\omega_{\rm cor} \to 0$ we have

$$\lim_{\omega_{\rm cor}\to\infty} G_1(\omega) = (\pi/2)^{1/2} = \text{const}$$

that is, the spectral density tends to a constant value, which corresponds to white noise or the delta-correlated process. Indeed, by performing the reverse Fourier transform, we obtain

$$G_1(\tau) = (2\pi)^{-1/2} \int_{-\infty}^{+\infty} d\omega \exp(-i\omega\tau) G_1(\omega) = (1/2)\tau_{\rm cor}^{-1} \exp(-|\tau|/\tau_{\rm cor})$$
$$\lim_{\tau_{\rm cor}\to 0} G_1(\tau) = \delta(\tau).$$

Substituting (5.3.22) and (5.3.33) into expression (5.3.20) and integrating over ω, we find

$$\alpha = \pi g_0 \int_0^\infty k^4 \, dk \, G_2(k) \frac{\omega_{\rm cor}^2}{(\nu_0 \omega_{\rm cor} + c^2)k^2 + \omega_{\rm cor}^2}. \tag{5.3.24}$$

We see from here that if we tend the integrand to the limit $\omega_{\rm cor} \to \infty$ (i.e. $\tau_{\rm cor} \to 0$), the result is

$$\alpha \sim \pi g_0 \int_0^\infty k^4 \, dk \, G_2(k) \sim H(0, 0) \sim \langle \boldsymbol{V}'(\boldsymbol{r}, t) \operatorname{rot} \boldsymbol{V}'(\boldsymbol{r}, t) \rangle$$

in agreement with the evaluation of (5.3.9). However, we will not perform the limiting transition yet and first calculate the integral over k. To do this, we make another assumption involving $G_2(k)$,

$$G_2(k) = \exp(-\lambda_{\rm cor}^2 k^2). \tag{5.3.25}$$

Substituting (5.3.25) into (5.3.24) and integrating, we obtain

$$\alpha = (1/4)\pi^{3/2} g_0 \lambda_{\rm cor}^{-5} \mu^2 (1 - 2\mu^2 + 2\pi^{1/2} \mu^3 \exp(\mu^2)[1 - \psi(\mu)])$$
$$\psi(\mu) = 2\pi^{-1/2} \int_0^\mu \exp(-t^2) \, dt \tag{5.3.26}$$

where we have introduced the dimensionless parameter μ,

$$\mu = \frac{\omega_{\rm cor}^2 \lambda_{\rm cor}^2}{\nu_0 \omega_{\rm cor} + c^2} = \frac{\lambda_{\rm cor}^2}{\nu_0 \tau_{\rm cor} + c^2 \tau_{\rm cor}^2}. \qquad (5.3.27)$$

Expression (5.3.26) is the exactly calculated value of α for the model correlation function $G(k, \omega)$, given by the relations (5.3.23) and (5.3.25). We will analyze it for two limiting cases, $\mu \gg 1$ and $\mu \ll 1$. In the first case, when $\mu \gg 1$, we will use the asymptotics of the error integral and obtain

$$\alpha \approx (3/8)\pi^{3/2} g_0 \lambda_{\rm cor}^{-5} (1 - (5/2)\mu^{-2}) \qquad \mu \gg 1. \qquad (5.3.28)$$

In the other limiting case , when $\mu \ll 1$, we have

$$\alpha \approx (1/4)\pi^{3/2} g_0 \lambda_{\rm cor}^{-5} (\mu^2 - 2\mu^4) \qquad \mu \ll 1.$$

For large values of μ, the main term of the asymptotic expansion is independent of the parameter μ and corresponds to the limiting case of a delta-correlated process. Indeed, expression (5.3.27) shows that as $\tau \to 0$ at finite values of the parameters ν_0 and c, $\mu \to \infty$.

In this limiting case $\alpha \sim H(0, 0)$, which is in agreement with the paper of Moiseev *et al* (1983a). If we neglect viscosity in the denominator of the fraction (5.3.27), assuming $\nu_0 \ll c^2 \tau_{\rm cor}$, we have $\mu^{1/2} \sim (\lambda_{\rm cor}/\tau_{\rm cor} c)$, an analogue to the Mach number. Obviously, the limit $\tau_{\rm cor} \to 0$, $\mu \to \infty$ corresponds formally to $c \to 0$, $\mu \to \infty$. For this reason, the speed of sound does not enter the resulting expression for delta-correlated processes. The transition to incompressible fluid is realized if the speed of sound tends to infinity. However, if $c \to \infty$, the compressibility parameter $\mu \to 0$ and at the limit we have $\alpha \to 0$. Therefore, we come to a familiar conclusion that there is no α-effect in the limiting case of the incompressible fluid.

5.3.5 Large-scale instability

The linearized equation (5.3.19) that describes vorticity has unstable solutions that describe generation of large-scale vortices. Indeed, changing to Fourier transforms and choosing $\boldsymbol{V}^{(1)}(\boldsymbol{k})$ as $\boldsymbol{V}^{(1)}(\boldsymbol{k}) = (V_x^{(1)}(\boldsymbol{k});\ V_y^{(1)}(\boldsymbol{k});\ 0)$, we find from (5.3.19) the instability increment

$$\gamma = \alpha k - \nu k^2$$

with the maximum value of the growth rate $\gamma_{\rm max} = \alpha^2/4\nu$, reached at $k = k_0 = \alpha/2\nu$, where ν is the turbulent viscosity coefficient proportional to turbulence energy, and the coefficient α is proportional to the helicity of turbulence. Therefore, the instability scale is $L = k_c^{-1}$. If the compressibility parameter is large ($\mu \gg 1$) the coefficient α is independent of μ and the instability scale is given by the relation

$$L \sim \frac{\langle (V')^2 \rangle}{\langle \boldsymbol{V}' \operatorname{rot} \boldsymbol{V}' \rangle} \sim \frac{\int \langle (V')^2 \rangle \, \mathrm{d}\boldsymbol{r}}{\int \langle \boldsymbol{V}' \operatorname{rot} \boldsymbol{V}' \rangle \, \mathrm{d}\boldsymbol{r}} \sim E/I$$

where I is the topological invariant, and E is the conventional invariant proportional to energy.

The instability scale increases at small values of μ, acquiring the factor μ^{-2}:

$$L \sim \mu^{-2} E/I \qquad \mu \ll 1.$$

The structures generated by helical turbulence are large scale only if $L \gg \lambda_{\mathrm{cor}}$, i.e. if $E/I \gg \lambda_{\mathrm{cor}}$, which is a constraint imposed on the correlator parameters. If $L \gg \lambda_{\mathrm{cor}}$, we then assume that $\langle (V')^2 \rangle \sim \lambda^2/\tau^2$ and obtain $T \sim \gamma^{-1} \gg \tau_{\mathrm{cor}}$, that is, the large-scale structure is automatically a slow structure.

One important factor must be emphasized. The structures generated are helical, that is, the scalar product in them is

$$(\boldsymbol{V}^{(1)} \operatorname{rot} \boldsymbol{V}^{(1)}) \neq 0.$$

5.3.6 Discussion

Since the averaged equation contains an α-term, it has an instability owing to the positive feedback between various components of the vector field which produces a large-scale vortex. We can ask a question: what is the role played by compressibility if we consider the excitation of a purely solenoidal field by solenoidal turbulence? To answer this question, we return to initial equations. The averaged equation (5.3.14) contains the quantity

$$\left\langle V_k' \frac{\partial V_i'}{\partial x_k} \right\rangle + \left\langle V_k' \frac{\partial V_i'}{\partial x_k} \right\rangle.$$

We rewrite it in the form

$$\left\langle V_k' \frac{\partial V_i'}{\partial x_k} \right\rangle + \left\langle V_k' \frac{\partial V_i'}{\partial x_k} \right\rangle = \frac{\partial}{\partial x_k} \left(\langle V_k' V_i' \rangle + \langle V_k' V_i' \rangle \right) - \langle V_i' \operatorname{div} \boldsymbol{V}' \rangle - \langle V_i' \operatorname{div} \boldsymbol{V}' \rangle. \tag{5.3.29}$$

The first term in the right-hand side of (5.3.29) cannot produce the α-term since the tensor $(\langle V_k' V_i' \rangle + \langle V_k' V_i' \rangle)$ is symmetrical in indices. The mean value vanishes, $\langle V_i \operatorname{div} \boldsymbol{V}' \rangle = 0$, since the turbulence correlator satisfies the incompressibility condition (see formula (5.3.8) and the comment to it). Therefore, the α-term appears because of the term $\langle V_i' \operatorname{div} \boldsymbol{V}' \rangle$. The scalar value $\varphi = \operatorname{div} \boldsymbol{V}'$ is described by the wave equation with the driving force,

$$\frac{\partial^2 \varphi}{\partial t^2} - c^2 \Delta \varphi - \nu_0 \Delta \frac{\partial \varphi}{\partial t} = -2 \frac{\partial}{\partial t} \left(\frac{\partial V_i^{(1)}}{\partial x_m} \right) \left(\frac{\partial}{\partial x_i} V_m' \right). \tag{5.3.30}$$

The right-hand side of equation (5.3.30) contains high frequency $\omega \sim 1/\tau_{\text{cor}}$ whose amplitude is modulated by a slowly-varying function $\boldsymbol{V}^{(1)}$. For this reason V' has the higher frequency of the same order of magnitude as $\boldsymbol{V}^t$. An approximate solution of this equation can be easily obtained if we perform the Fourier transform in rapidly-varying variables:

$$\varphi(\boldsymbol{k}, \omega) \cong -2\frac{\partial V_i^{(1)}}{\partial x_m} \frac{\omega k_i V_m^t(\boldsymbol{k}, \omega)}{c^2k^2 - \omega^2 + \mathrm{i}\nu_0 k^2\omega}. \tag{5.3.31}$$

Multiplying this equality by V_j^t and performing the averaging, we find

$$\langle V_j^t(\boldsymbol{k}_1, \omega_1)\varphi(\boldsymbol{k}, \omega)\rangle = -2\frac{\partial V_i^{(1)}}{\partial x_m} \frac{\omega k_i \langle V_j^t(\boldsymbol{k}_1, \omega_1) V_m^t(\boldsymbol{k}, \omega)\rangle}{c^2k^2 - \omega^2 + \mathrm{i}\nu_0 k^2\omega}. \tag{5.3.32}$$

Converting Fourier transforms back to the coordinate representation, we find for the one-point average

$$\begin{aligned}\langle V_j^t(\boldsymbol{r}, t)\varphi(\boldsymbol{r}, t)\rangle &= -2\frac{\partial V_i^{(1)}}{\partial x_m}(\boldsymbol{r}, t) \int \exp\left\{\mathrm{i}(\boldsymbol{k} + \boldsymbol{k}_1)\boldsymbol{r} - \mathrm{i}(\omega + \omega_1)\tau\right\} \\ &\quad \times \frac{\omega k_i \langle V_j^t(\boldsymbol{k}_1, \omega_1) V_m^t(\boldsymbol{k}_1, \omega)\rangle}{c^2k^2 - \omega^2 + \mathrm{i}\nu k^2\omega} \,\mathrm{d}\boldsymbol{k}_1 \,\mathrm{d}\boldsymbol{k} \,\mathrm{d}\omega_1 \,\mathrm{d}\omega.\end{aligned} \tag{5.3.33}$$

Substituting here the correlator as a Fourier transform,

$$\begin{aligned}\langle V_j^t(\boldsymbol{k}_1, \omega_1) V_m^t(\boldsymbol{k}, \omega)\rangle &= \left\{D(k, \omega)\left(\delta_{jm} - \frac{k_j k_m}{k^2}\right) - \mathrm{i}G(k, \omega)\varepsilon_{jml}k_l\right\} \\ &\quad \times \delta(\boldsymbol{k}_1 + \boldsymbol{k})\,\delta(\omega_1 + \omega)\end{aligned} \tag{5.3.34}$$

and recalling that $\varphi = \operatorname{div} \boldsymbol{V}'$, we arrive at the expression

$$\langle V_j^t(\boldsymbol{r}, t) \operatorname{div} \boldsymbol{V}'(\boldsymbol{r}, t)\rangle = -2\varepsilon_{jmn}\frac{\partial V_i^{(1)}}{\partial x_m}(\boldsymbol{r}, t) \int \frac{\mathrm{i}\omega k_i k_n G(\boldsymbol{k}, \omega)}{c^2k^2 - \omega^2 + \mathrm{i}\nu_0 k^2\omega} \,\mathrm{d}\boldsymbol{k} \,\mathrm{d}\omega \tag{5.3.35}$$

which gives, after integration over angles,

$$\langle V_j^t(\boldsymbol{r}, t) \operatorname{div} \boldsymbol{V}'(\boldsymbol{r}, t)\rangle = \alpha \operatorname{rot} \boldsymbol{V}^{(1)}(\boldsymbol{r}, t) \tag{5.3.36}$$

where α is given by (5.3.26). Therefore, the α-term in the averaged equation arises owing to the mean value $\langle \boldsymbol{V}^t \operatorname{div} \boldsymbol{V}'\rangle$, in other words, owing to the arising inhomogeneity scalar field $\operatorname{div} \boldsymbol{V}'$. The quantity $\alpha \operatorname{rot} \boldsymbol{V}$ is a slowly-varying one, and therefore the rapidly-varying $\operatorname{div} \boldsymbol{V}'$ is compensated for by the rapid component $\boldsymbol{V}^t$ which is always present in the turbulence's continuous spectrum.

Note that $1/T \ll \omega_{\text{turb}}$, where T is the characteristic time of changes in the averaged field and $\omega_{\text{turb}} \sim 1/T_{\text{cor}}$ is the characteristic turbulence frequency.

Therefore, the two-scale approach also assumes the turbulent velocity field to be rapidly-varying.

The two factors contributing to integral (5.3.35) are, first, the peculiarities of the correlation function, and second, the poles of the denominator. Note that for a model correlation function given by expressions (5.3.23), these contributions are of the same order of magnitude. If we consider a δ-correlated process, the function $G(\omega) = \text{const}$ has no singularities; hence only those of the denominator contribute to the integral over ω; note that this contribution is independent of the position of the pole and is thus independent of the speed of sound c.

The α-term in the averaged equation results in a positive feedback between various components of the solenoidal field. The appearance of the α-term stems not from compressibility but mostly from the turbulence's helicity. The role of compressibility is that of an additional factor that modifies the symmetry of equations.

Helicity is one of the most important characteristics of any vector field. The integral over a fluid's volume, $I = \int (\boldsymbol{V} \operatorname{rot} \boldsymbol{V})\, \mathrm{d}^3\boldsymbol{r}$, is an integral of motion in the ideal fluid and thus characterizes the topology of the vector field $\boldsymbol{V}$, determining the degree of linkage of the streamlines in the fluid. Note that the invariant I is a pseudoscalar and therefore, motions with $I \neq 0$ are not mirror invariant.

5.4 Large-scale vortical instability and convection: further progress

5.4.1 Equation of vortical instability due to convective turbulence and Coriolis force

We will consider this topic following the work of Rutkevich (1993). As follows from the previous sections, the concept of helical turbulence allows us to derive a set of averaged equations that describe the generation of large-scale vortices of the type of tropical cyclones. This model typically involved linked streamlines of the toroidal and poloidal velocity fields in the fluid; a feedback between these fields causes the generation of a large-scale vortex. The helicity, that is, the violation of the mirror invariance of turbulence, implies, even if implicitly, the presence of a certain convective process with a Coriolis force as its background (Vainshtein *et al* 1980). This concept proves to be so efficient for sustaining the regime of structure generation that it needs no explicit support of such obviously important factors as the Coriolis force and convection on the energy-containing scale. By interpreting large-scale vortices obtained within this model, for example, the initial stage of formation of tropical cyclones, we have to assume that the concept of turbulence helicity comprises within itself, or parametrizes, all these factors. However, the need to include helicity into the model phenomenologically remains its shortcoming.

Berezin and Zhukov (1989) made an attempt to construct a model of generation of large-scale hydrodynamic structures directly by the Coriolis force. As a result, they came to an erroneous conclusion that such a model cannot be constructed for an incompressible fluid; for compressible fluids, they derived equations for large-scale motions against the background of a strong vertical inhomogeneity and planned to analyze them numerically. Consequently, this work left unanswered the question of the possibility of constructing a model of generation of large-scale hydrodynamic structures in incompressible fluids against the background of small-scale convection in the field of the Coriolis force.

The turbulence arising in the area of a tropical depression is usually of convective nature and, according to the majority of models, is maintained by the release of the latent heat of vapor condensation. This process corresponds best to the problem of instability created in a fluid containing an internal heating source. The temperature profile under these conditions is a quadratic parabola whose curvature is proportional to the intensity of the heat source. According to the fundamental concepts of helical turbulence (Vainshtein *et al* 1980, Moiseev *et al* 1988a, Krause and Radler 1980), the procedure for calculating the Reynolds stress by statistical averaging over small-scale turbulent pulsations, taking into account the Coriolis force and the curvature of the temperature profile in a convective cell, must lead to averaged equations that are analogs of the equations of generation of large-scale vortices by helical turbulence (see the previous sections of this chapter). We emphasize that taking into account the vertical inhomogeneity within a simple problem of convection on a small scale with constant vertical temperature gradient is not sufficient for deriving adequate Reynolds stress.

When modeling convective turbulence, we assume that its characteristic scale does not exceed the size of convective cells. We also assume that the local convective processes are caused by local heat released within a λ-thick layer, with insignificant stable stratification existing above and below this layer. The temperature profile in the convective layer will be chosen as the Taylor series expansion in the vertical coordinate z, with the profile curvature assumed to be small:

$$T_0(z) = \text{const} - Az - \frac{B}{2}z^2 + \ldots, \qquad A, B > 0 \qquad A \gg \frac{B}{2}\lambda. \quad (5.4.1)$$

The behavior of convection under such conditions was analyzed by Gershuni and Zhukhovitskii (1972). The critical Reynolds number in this system is found to be considerably smaller than the Reynolds number for the convection in the bounded fluid layer; owing to penetrating convection, the cells get vertically elongated.

To analyze the derivation of equations for large-scale instability in incompressible fluids caused by small-scale convective turbulence with the Coriolis force as its background, we will use the following set of equations

(Gershuni and Zhukhovitskii 1972):

$$\begin{aligned}
&\frac{\partial V_i}{\partial t} - \nu\Delta V_i + V_k\nabla_k V_i + \frac{\nabla_i P}{\rho} + g e_i + 2\Omega\varepsilon_{ijs}e_j V_s = 0 \\
&\frac{\partial T}{\partial t} - \chi\Delta T + V_k\nabla_k T = Q \\
&\nabla_k V_k = 0 \\
&\nabla_k = \frac{\partial}{\partial x_k} \\
&\rho = \rho_0(1 - \beta T).
\end{aligned} \tag{5.4.2}$$

Here ν is the kinematic viscosity, χ is the temperature conductance (we assume the Prandtl number to be unity, $\mathrm{Pr} = \nu/\chi = 1$),

$$\beta = -\frac{1}{\rho}\left(\frac{\partial \rho}{\partial T}\right)_p$$

is the thermal expansion coefficient, T is temperature, $\boldsymbol{e}$ is the unit vector pointing vertically upward, and Ω is the angular velocity.

We assume that a layer with unstable stratification and temperature profile (5.4.1) arises in the fluid at some time, caused by an internal heat source Q against the background of weak stable stratification. We will conduct instability analyses for perturbations in velocity $\boldsymbol{V}(t, x)$, temperature $\theta(t, x)$ ($T = T_0(z) + \theta(t, \boldsymbol{x})$) and pressure $P_1(t, \boldsymbol{x})$ ($P = P_0(z) + P_1(t, \boldsymbol{x})$, grad $P_0(z) = \rho_0 g \boldsymbol{e}(1 - \beta T_0(z))$), against the background of the ground state $T_0(z)$ and $P_0(z)$ due to heating. The set of equations for perturbations in the Boussinesq approximation takes the form

$$\frac{\partial V_i}{\partial t} - \nu\Delta V_i + V_k\nabla_k V_i + \frac{\nabla_i P_1}{\rho_0} - g e_i \beta\theta + 2\Omega\varepsilon_{isj}e_s V_j = F_i \tag{5.4.3}$$

$$\frac{\partial \theta}{\partial t} - \chi\Delta\theta - (V_k e_k)(A + Bz) + V_k\nabla_k\theta = 0. \tag{5.4.4}$$

A random external force F_i ($\langle F_i\rangle = 0$) was added to the Navier–Stokes equation (5.4.3), which results in a small-scale turbulence in the fluid; we assume its parameters to be synchronized with those of the convective process. This means, among other things, that the homogeneous steady-state turbulence must be regarded as anisotropic. The correlator of a random anisotropic velocity field in the Fourier transform representation in coordinates is well known (see, for example, Vainshtein *et al* 1980):

$$\begin{aligned}
T_{ij}(t_1 - t_2, \boldsymbol{k}) = G(t_1 - t_2, \boldsymbol{k}_\perp, \boldsymbol{ek})\Bigg\{&\left(\delta_{ij} - (1-\mu)\frac{k_i k_j}{k^2}\right) \\
&+ \mu\left(e_i e_j \frac{k^2}{(\boldsymbol{ek})^2} - \frac{e_i k_j + e_j k_i}{(\boldsymbol{ek})}\right)\Bigg\}
\end{aligned}$$

where μ is a parameter that characterizes the degree of anisotropy ($0 \leqslant \mu \leqslant 1$).

In this context we limit the analysis to deriving the Reynolds stress responsible for large-scale instability, with the Coriolis force and small-scale penetrating convection against the background of non-helical anisotropic turbulence possessing very simple properties. The Reynolds stress that produces the turbulence correction to the viscosity coefficient was investigated by Krause and Rudiger (1974). The characteristic turbulence scale is of the order of the thickness λ of the layer with unstable stratification, and the turbulence intensity u_T is determined by the power of the external heat source.

This model assumes that the conditions for the appearance of small-scale convection are satisfied; this means that the steady-state part of the small-scale linear operator vanishes. The weak increment of convective instability $\gamma \ll \nu/\lambda^2$ is assumed to be consistent with the relatively large turbulence correlation time $\tau = 1/\gamma$. In fact, this means that the convective motions are treated as turbulence. We could say that the process of free convection, which generates turbulent motions in the field of an inhomogeneous temperature gradient and Coriolis force, is at the same time an efficient mechanism for generating the helicity of this turbulence. However, its analysis can be carried out with comparative ease within the framework of the conventional scheme for Reynolds stress calculation, provided the turbulent correlation time is still not too long:

$$1 \gg \frac{\lambda^2}{\nu\tau} \gg \frac{g\beta B\lambda^2}{2\nu^2} = \frac{\mathrm{Re}}{2}\frac{B\lambda}{A} \simeq \frac{B\lambda}{A}.$$

The Reynolds stress formula is derived on the basis of equations (5.4.2)–(5.4.4), with the nonlinear terms in equations (5.4.3) and (5.4.4) assumed to be small (i.e. the Reynolds number $\mathrm{Re} = u_T\lambda/\nu \ll 1$). It is convenient to rewrite the set (5.4.3) and (5.4.4) as a single equation for velocity, solving the equation for temperature perturbations (5.4.3) by iterations:

$$L_{ij}DV_j - \beta g P_{im}e_m\frac{1}{D}(Bz)e_jV_j = F_i - P_{im}\nabla_k(V_kV_m) - \beta A g\frac{1}{D}P_{im}e_me_j\nabla_k\left(V_k\frac{1}{D}V_j\right). \tag{5.4.5}$$

Here we have introduced the notation L_{ij} for the coordinate-independent part of the linear operator:

$$L_{ij} = \delta_{ij} - qP_{im}e_me_j + bP_{im}\varepsilon_{msj}e_s$$

$$D = \frac{\partial}{\partial t} - \nu\Delta$$

$$q = \frac{\beta g A}{D^2}$$

$$b = \frac{2\Omega}{D}$$

$$P_{im} = \delta_{im} - \frac{\nabla_i\nabla_m}{\Delta}.$$

The projection operator P_{im} excludes the potential component of the velocity field. The differential operators in the denominator are interpreted as integral operators with the corresponding Green's functions.

Note that in general the operator L_{ij} is a function of the large-scale velocity that is a result of the evolution of the instability we seek here. Therefore, it describes the phenomenon of driven convection, and taking it into account could describe the reverse effect of instability on the formation of helicity. In this model, however, it is not advisable to include the large-scale velocity into the operator L_{ij} because we neglect the changes in large-scale velocity fields on a small scale, so that taking the large-scale velocity into account in the operator L_{ij} will only describe the Galilean translation of a small-scale cell as a whole and will not affect the parameters of the convective process.

For the averaging procedure of equation (5.4.5) we follow Krause and Rudiger (1974) and recast the velocity field V_i to the form

$$V_i = V_i^T + \tilde{V}_i + \langle V_i \rangle.$$

The field V_i^T is a homogeneous anisotropic velocity field due directly to the action of the external force F_i:

$$\begin{aligned} L_{ij} D V_j^T - \beta g P_{im} e_m \frac{1}{D}(Bz) e_j V_j^T \\ = F_i - P_{im}(V_k^T V_m^T) - \beta A g \frac{1}{D} P_{im} e_m e_j \nabla_k \left(V_k^T \frac{1}{D} V_j^T \right). \end{aligned} \quad (5.4.6)$$

The quantity $\langle V_i \rangle$ is the mean large-scale velocity field $\langle V \rangle \ll V^T$. The equation for it is obtained by averaging over the ensemble of realizations of equation (5.4.5):

$$L_{ij} D \langle V_j \rangle = -P_{im} \nabla_k \left(\langle V_k^T \tilde{V}_m \rangle + \langle \tilde{V}_k V_m^T \rangle \right). \quad (5.4.7)$$

When deriving (5.4.7), we make use of the fact that the large-scale background temperature profile is assumed to be neutral on average.

The mean velocity field $\langle V_i \rangle$ against the background of turbulent pulsations V_i^T generates a small inhomogeneous additional term $\tilde{V}_i$ ($\tilde{V} \ll V_T$) which is, therefore, a functional of V^T and $\langle V \rangle$: $\tilde{V}_i = \tilde{V}_i\{V^T, \langle V \rangle\}$. Subtracting the equation for the turbulent part of the velocity V_i^T (5.4.6) and the averaged equation (5.4.7) from the total equation (5.4.5), we obtain in the lowest order an equation that describes the inhomogeneous part of the turbulent velocity $\tilde{V}_i$:

$$\begin{aligned} L_{ij} D \tilde{V}_j - \beta g P_{im} e_m \frac{1}{D}(Bz) e_j \tilde{V}_j = -P_{im} \nabla_k (V_k^T \langle V_m \rangle + \langle V_k \rangle V_m^T) \\ - \beta A g \frac{1}{D} P_{im} e_m e_j \nabla_k \left(V_k^T \frac{1}{D} \langle V_j \rangle + \langle V_k \rangle \frac{1}{D} V_j^T \right). \end{aligned}$$

Equation (5.4.7) for the large-scale velocity includes the mean values of the quadratic combinations (Reynolds stress). They can be expressed in terms of the mean field $\langle V_i \rangle$ and a turbulence correlator, using the functional dependence of the field $\tilde{V}_i$ on the random field V_i^T, which we consider to be Gaussian (Furutsu–Novikov formula (Klyatskin 1980)):

$$\langle V_k^T(t, \boldsymbol{x}) \tilde{V}_m(t, \boldsymbol{x}) \rangle = \lim_{\substack{t_1 \to t \\ \boldsymbol{x}_t \to \boldsymbol{x}}} \int \mathrm{d}s \int \mathrm{d}\boldsymbol{y} \, \langle V_k^T(t, \boldsymbol{x}) V_r^T(s, \boldsymbol{y}) \rangle \left\langle \frac{\delta \tilde{V}_m(t_1, \boldsymbol{x}_1)}{\delta V_r^T(s, \boldsymbol{y})} \right\rangle$$

by analogy to what was done when analyzing the generation of large-scale structures against the background of helical turbulence (see the preceding sections of this chapter). Therefore, the subsequent manipulations with equation (5.4.7) are similar to those in the case of helical turbulence. We refer the reader for the details to Rutkevich (1993) but will present certain features of the research conducted. It must be emphasized again that turbulence is treated as non-helical but the Coriolis force is taken into account explicitly. The main attention is then paid to that part of the additional inhomogeneous term in the turbulent velocity which stems from the non-uniform temperature gradient. In the case of strong anisotropy of turbulence ($\mu \to 1$), the energy is predominantly concentrated in vertical motions while the coefficient G_0, which characterizes the value of the correlation function, is related to the turbulence energy density in the following way: $G_0 = (2\pi)^2 E\eta^2$.

The equation for the mean velocity can be written as follows:

$$\left(\frac{\partial}{\partial t} - \nu_T \Delta\right) \langle V_i \rangle + 2\Omega P_{im} \varepsilon_{msj} e_s \langle V_i \rangle = C P_{im} \nabla_k \Big[- \big(\langle V_k^{\perp} \rangle e_m + \langle V_m^{\perp} \rangle e_k\big)$$
$$- 2 e_k e_m (e_s \langle V_s \rangle) + \tfrac{1}{2} b^* (e_k \varepsilon_{mrs} + e_m \varepsilon_{krs}) e_r \langle V_s \rangle \Big] \qquad (5.4.8)$$

where ν_T is the turbulent viscosity coefficient on a large scale; its value is much larger than that of the corresponding small-scale coefficient ν ($\nu_T \gg \nu$) and

$$C = E\eta^2 \left(\frac{\nu\tau}{\lambda^2}\right)^3 \frac{B\lambda}{A} \frac{\lambda}{\nu} \qquad b^* = \frac{2\Omega\lambda^2}{\nu}.$$

Equation (5.4.8) describes the large-scale instability in an incompressible fluid. This instability is conveniently analyzed in terms of the poloidal and toroidal velocity fields $\langle \boldsymbol{V} \rangle$:

$$\langle \boldsymbol{V} \rangle = \mathrm{rot}(\boldsymbol{e}\psi) + \mathrm{rot}\,\mathrm{rot}(\boldsymbol{e}\phi).$$

The set of equations for the potential poloidal ϕ and toroidal ψ velocity fields is of the form

$$\left(\frac{\partial}{\partial t} - \nu_T \Delta\right) \Delta\phi + 2\Omega \boldsymbol{e}\nabla\psi$$
$$= C \left[(3\Delta_{\perp} - (\boldsymbol{e}\nabla)^2) \boldsymbol{e}\nabla\phi + \frac{b^*}{2}((\boldsymbol{e}\nabla)^2 - \Delta_{\perp})\psi \right] \qquad (5.4.9)$$
$$\left(\frac{\partial}{\partial t} - \nu_T \Delta\right) \Delta\psi + 2\Omega \boldsymbol{e}\nabla\phi = -C \left[\boldsymbol{e}\nabla\psi + \frac{b^*}{2}(\boldsymbol{e}\nabla)^2\phi \right].$$

The large-scale instability described by equations (5.4.9) is best manifest for structures of horizontal size larger than the vertical size, $K_z \gg K_\perp$, where $\boldsymbol{K} = (K_z, K_\perp)$ is the wave vector of large-scale instability, whose increment is determined by

$$\gamma = \pm C\frac{b^*}{2}K_z - \nu_T K_z^2. \tag{5.4.10}$$

This results nullifies the conclusion drawn by Berezin and Zhukov (1989) on the impossibility of this instability in incompressible fluids. The evolution of the instability generates structures of characteristic size $K_{zm}^{-1} = 4\nu_T/Cb^*$ that realizes the maximum of the increment (5.4.10); the frequencies of the forced oscillations of the medium are $\omega = -CK_z \pm 2\Omega$.

Consider a limiting case of large values of the parameter $b^* = 2\Omega\lambda^2/\nu \gg 1$ which is realized, for example, at sufficiently small values of the small-scale viscosity coefficient ν. Equation (5.4.8), with lengths in units of the layer thickness h and times in units of h^2/ν_T, takes the form

$$\left(\frac{\partial}{\partial t} - \nu\Delta\right)\langle V_i\rangle + \frac{2\Omega h^2}{\nu_T} P_{im}\varepsilon_{msj}e_s\langle V_i\rangle$$
$$= E\frac{\lambda h}{\nu\nu_T}\eta^2\left(\frac{\nu\tau}{\lambda^2}\right)^3 \frac{B\lambda}{A}\frac{2\Omega\lambda^2}{\nu} P_{im}\nabla_k(e_k\varepsilon_{mrs} + e_m\varepsilon_{krs})e_r\langle V_s\rangle. \tag{5.4.11}$$

Equation (5.4.11) differs from the equation of large-scale instability derived by Lupyan *et al* (1992) on the basis of the concept of helical turbulence only in a term with the Taylor number that finds a natural place in this model. The role of the helicity coefficient S is therefore played by the product

$$S = 2\Omega\tau\eta^2\frac{E\tau^2}{\lambda^2}\frac{h\nu}{\lambda\nu_T}\frac{B\lambda}{A}.$$

We thus find that in this model the procedure of statistical averaging over small-scale turbulent pulsations (taking into account the Coriolis force and the curvature of the temperature profile) leads to large-scale equations that are similar to the equations in the model of generation of large-scale vortices by helical turbulence (Moiseev *et al* 1983b, 1987, 1988a; Lupyan *et al* 1992, 1993). This means that the concept of helical turbulence parametrizes the joint action of inhomogeneity of the temperature gradient and the Coriolis force and that the helicity coefficient is proportional to the product of the inhomogeneity of the temperature gradient, turbulence intensity and the Coriolis force.

Note that for both sets (5.4.9) and (5.4.11), the large-scale instability is caused by interaction of the poloidal and toroidal velocity fields; as in the case of helical turbulence, its increment is independent of the sign of the helicity parameters $Cb^*/2$ or S. The sign of the helicity characterizes the mutual directions of the velocity fields; with the direction of, say, the poloidal field fixed (for example, the fluid rises in the central part of the cell and sinks at the boundaries), the sign of the helicity parameter controls the sign of the toroidal field, that is, the direction of rotation of a large-scale vortex (if the helicity sign is positive, the vortex rotates counterclockwise (as seen from above), and if the sign is negative, the rotation is clockwise).

If the helicity coefficient S exceeds a certain threshold level, equation (5.4.11) describes the generation of a large-scale tropical-cyclone-type vortex whose horizontal size is much greater than the layer thickness h. Under special conditions, the toroidal velocity field in this vortex is much stronger than the poloidal field. Taking into account the term with the Taylor number does not lead to a qualitative change in the instability process. Lupyan *et al* (1993) showed that the role played by this term reduces to a somewhat increased instability threshold and to changes in the shape of neutral curves.

5.4.2 Rayleigh–Bénar convection in turbulent fluid at very large Reynolds numbers

Helicity was investigated in the preceding sections at low Reynolds numbers. We will demonstrate now that this constraint is not of principal importance.[2] In what follows we consider the dynamics of highly turbulent fluid (at very high Reynolds numbers), heated from below in the field of gravity. We assume here, just as we did earlier, that small-scale helical turbulence plays a predominant part in the formation of large-scale structures driven by the vortical dynamo mechanism.

The interaction between large-scale structures and small-scale turbulence is studied using the two-scale direct interactions approximation (TSDIA) (Yoshizawa 1992, Yoshizawa and Yokoi 1991, Yokoi and Yoshizawa 1993, Leslie 1973). We will show that if the small-scale turbulence has no mirror symmetry, and the total helicity $\alpha \propto \int \langle \boldsymbol{v}^T \boldsymbol{\Omega}^T \rangle \, \mathrm{d}\boldsymbol{x}$, where $\boldsymbol{\Omega}^T = \nabla \times \boldsymbol{v}^T$, is above a certain threshold value, a single large-scale vortex may be formed instead of a multitude of convective cells.

Let us consider an incompressible turbulent fluid heated from below in the field of gravity. The dynamics of the incompressible fluid are then described by the equations of motion and heat conduction in the Boussinesq approximation,

$$\frac{\partial \boldsymbol{v}}{\partial t} + (\boldsymbol{v}\nabla)\boldsymbol{v} = -\nabla p + \nu_0 \nabla^2 \boldsymbol{v} + \boldsymbol{g}\beta\theta \qquad (5.4.12)$$

$$\frac{\partial \theta}{\partial t} + (\boldsymbol{v}\nabla)\theta = \chi_0 \nabla^2 \theta + \boldsymbol{\kappa}\cdot\boldsymbol{v} \qquad (5.4.13)$$

and the condition of incompressibility,

$$\nabla\cdot\boldsymbol{v} = 0 \qquad (5.4.14)$$

[2] This section is based on the results obtained by Moiseev together with Mikhailenko and Fedutenko.

where $\boldsymbol{v}$ and θ are the velocity and temperature perturbations, respectively, p is pressure divided by the density of the fluid, ν_0 and χ_0 are the molecular viscosity and heat conductivity, respectively, $\boldsymbol{g}$ is the gravitational acceleration, $\boldsymbol{\kappa} = -\nabla T_0$ is the temperature gradient with the sign reversed and assumed to be known, and β is the bulk expansion coefficient. We are interested in the structures forming in the large-scale flow, averaged over the ensemble of realizations. Separating the parameters of the fluid $\boldsymbol{f}$ into the component averaged over the ensemble $\langle \boldsymbol{f} \rangle$ and the fluctuational component $\boldsymbol{f}^T$, that is, $\boldsymbol{f} = \langle \boldsymbol{f} \rangle + \boldsymbol{f}^T$, $\langle \boldsymbol{f} \rangle = 0$, $\boldsymbol{f} = (\boldsymbol{v}, p, \theta)$, $\langle \boldsymbol{f} \rangle = (\langle \boldsymbol{v} \rangle, \langle p \rangle, \langle \theta \rangle)$, $\boldsymbol{f}^T = (\boldsymbol{v}^T, p^T, \theta^T)$, and substituting these expressions into (5.4.12)–(5.4.14), we arrive at a set of equations for the mean and the pulsational components of the flux,

$$\frac{\partial \langle \boldsymbol{v} \rangle}{\partial t} + (\langle \boldsymbol{v} \rangle \nabla) \langle \boldsymbol{v} \rangle = -\nabla \langle p \rangle + \nu_0 \nabla^2 \langle \boldsymbol{v} \rangle + \boldsymbol{g}\beta \langle \theta \rangle - \boldsymbol{F}^{(R)} \tag{5.4.15}$$

$$\frac{\partial \langle \theta \rangle}{\partial t} + (\langle \boldsymbol{v} \rangle \nabla) \langle \theta \rangle = \chi_0 \nabla^2 \langle \theta \rangle + \boldsymbol{\kappa} \cdot \langle \boldsymbol{v} \rangle - \nabla \cdot \boldsymbol{F}^{(\theta)} \tag{5.4.16}$$

$$\nabla \cdot \langle \boldsymbol{v} \rangle = 0$$

$$\frac{\partial \boldsymbol{v}^T}{\partial t} + (\langle \boldsymbol{v} \rangle \nabla) \boldsymbol{v}^T + (\boldsymbol{v}^T \nabla) \langle \boldsymbol{v} \rangle + \left[(\boldsymbol{v}^T \nabla) \boldsymbol{v}^T - \boldsymbol{F}^{(R)} \right]$$
$$= -\nabla p^T + \nu_0 \nabla^2 \boldsymbol{v}^T + \boldsymbol{g}\beta\theta^T \tag{5.4.17}$$

$$\frac{\partial \theta^T}{\partial t} + (\boldsymbol{v}^T \nabla) \langle \theta \rangle + (\langle \boldsymbol{v} \rangle \nabla) \theta^T = \chi_0 \nabla^2 \theta^T + \boldsymbol{\kappa} \cdot \boldsymbol{v}^T + \nabla \cdot \boldsymbol{F}^{(\theta)}$$

$$\nabla \cdot \boldsymbol{v}^T = 0 \tag{5.4.18}$$

where $\boldsymbol{F}^{(R)} = \langle (\boldsymbol{v}^T \nabla) \boldsymbol{v}^T \rangle$ and $\boldsymbol{F}^{(\theta)} = \langle \boldsymbol{v}^T \theta^T \rangle$.

The conditions of incompressibility (5.4.18) applied to (5.4.17) result in the Poisson equation for p^T:

$$\nabla^2 p^T = \beta (\boldsymbol{g} \nabla) \theta^T - \nabla \left[(\boldsymbol{v}^T \nabla) \boldsymbol{v}^T + \boldsymbol{F}^{(R)} \right] - 2 \frac{\partial \langle v_i \rangle}{\partial x_j} \frac{\partial v_j^T}{\partial x_i}$$

(we assume summation over repeated indices).

In the spirit of TSDIA, we introduce small-scale variables $\boldsymbol{x}$ and large-scale variables $\boldsymbol{X}$, and also display a number of relations yielded by the analysis cited: $\boldsymbol{f} = \langle \boldsymbol{f}(\boldsymbol{X}, t) \rangle + \boldsymbol{f}^T(\boldsymbol{x}, \boldsymbol{X}, t)$,

$$\boldsymbol{f}^T(\boldsymbol{x}, \boldsymbol{X}; t) = \frac{1}{(2\pi)^6} \int \tilde{\boldsymbol{f}}(\boldsymbol{k}, \boldsymbol{q}; t) \exp(-\mathrm{i}\boldsymbol{k} \cdot \boldsymbol{x} - \mathrm{i}\boldsymbol{q} \cdot \boldsymbol{X}) \, \mathrm{d}\boldsymbol{k} \, \mathrm{d}\boldsymbol{q}$$

$$\langle \boldsymbol{f}(\boldsymbol{X}; t) \rangle = \frac{1}{(2\pi)^3} \int \langle \boldsymbol{f}(\boldsymbol{q}; t) \rangle \exp(-\mathrm{i}\boldsymbol{q} \cdot \boldsymbol{X}) \, \mathrm{d}\boldsymbol{q}$$

$$\boldsymbol{v}^T(\boldsymbol{k}, \boldsymbol{q}; t) \approx \boldsymbol{v}^{T(0)}(\boldsymbol{k}; t) + \boldsymbol{v}^{T(1)}(\boldsymbol{k}, \boldsymbol{q}; t)$$

$$\theta^T(\boldsymbol{k}, \boldsymbol{q}; t) \approx \theta^{T(0)}(\boldsymbol{k}; t) + \theta^{T(1)}(\boldsymbol{k}, \boldsymbol{q}; t)$$

$$\theta^{T(0)}(\boldsymbol{k}; t) = \kappa_i \int_{-\infty}^{t} \mathrm{d}t_1 \, \hat{F}(\boldsymbol{k}; t, t_1) v_i^{T(1)}(\boldsymbol{k}; t_1)$$

$$\theta^{T(1)}(\boldsymbol{k},\boldsymbol{q};t) = \kappa_i \int_{-\infty}^{t} \mathrm{d}t_1\, \hat{F}(\boldsymbol{k};t,t_1) v_i^{T(1)}(\boldsymbol{k};t_1) + \mathrm{i}q_i \int_{-\infty}^{t} \mathrm{d}t_1 \hat{F}(\boldsymbol{k};t,t_1)\langle\theta(\boldsymbol{q};t_1)\rangle v_i^{T(0)}(\boldsymbol{k};t_1) + \mathrm{i}q_i \int_{-\infty}^{t} \mathrm{d}t_1 \hat{F}(\boldsymbol{k};t,t_1)\theta^{T(0)}(\boldsymbol{q};t_1)\langle v_i(\boldsymbol{k};t_1)\rangle$$

where $\hat{F}(\boldsymbol{k};t,t_1)$ are Green's functions of equation (5.4.16):

$$\left(\frac{\partial}{\partial t} + \chi_0 k^2\right)\hat{F}(\boldsymbol{k};t,t_1) + \mathrm{i}\int \mathrm{d}\boldsymbol{k}_1\,\mathrm{d}\boldsymbol{k}_2\,\delta(\boldsymbol{k}-\boldsymbol{k}_1-\boldsymbol{k}_2) \times k_{2i} v_i^{T(0)}(\boldsymbol{k}_1;t_1)\hat{F}(\boldsymbol{k}_2;t,t_1) = \delta(t-t_1)$$

$$\left(\frac{\partial}{\partial t} + \nu_0 k^2\right) v_i^{T(0)}(\boldsymbol{k};t) - \mathrm{i}M_{ijl}\int \mathrm{d}\boldsymbol{k}_1\,\mathrm{d}\boldsymbol{k}_2\,\delta(\boldsymbol{k}-\boldsymbol{k}_1-\boldsymbol{k}_2) v_j^{T(0)}(\boldsymbol{k}_1;t_1) v_l^{T(0)}(\boldsymbol{k}_2;t_1) - D_{ij}(\boldsymbol{k})\kappa_l g_j \beta \int_{-\infty}^{T} \mathrm{d}t_1\, \hat{F}(\boldsymbol{k}_2;t,t_1) v_l^{T(0)}(\boldsymbol{k}_1;t_1) = 0 \tag{5.4.19}$$

$$\left(\frac{\partial}{\partial t} + \nu_0 k^2\right) v_i^{T(1)}(\boldsymbol{k};t) - 2\mathrm{i}M_{ijl}\int \mathrm{d}\boldsymbol{k}_1\,\mathrm{d}\boldsymbol{k}_2\,\delta(\boldsymbol{k}-\boldsymbol{k}_1-\boldsymbol{k}_2) \times v_j^{T(0)}(\boldsymbol{k}_1;t_1) v_l^{T(1)}(\boldsymbol{k}_2;t_1) - D_{ij}(\boldsymbol{k})\kappa_l g_j \beta \int_{-\infty}^{T} \mathrm{d}t_1\, \hat{F}(\boldsymbol{k};t.t_1) v_l^{T(1)}(\boldsymbol{k};t_1)$$
$$= \mathrm{i}D_{ij}^{(1)}(\boldsymbol{k}) q_l \langle v_j(\boldsymbol{q};t)\rangle v_l^{T(0)}(\boldsymbol{k};t) + \mathrm{i}k_j\langle v_j(\boldsymbol{q};t)\rangle v_l^{T(0)}(\boldsymbol{k};t) + \mathrm{i}D_{ij}(\boldsymbol{k}) q_l q_j \beta \int_{-\infty}^{T} \mathrm{d}t_1\, \hat{F}(\boldsymbol{k}_2;t,t_1)\langle\theta(\boldsymbol{q};t_1)\rangle v_l^{T(0)}(\boldsymbol{k};t_1) \tag{5.4.20}$$

where

$$D_{ij}(\boldsymbol{k}) = \delta_{ij} - \frac{k_i k_j}{k^2} \qquad D_{ij}^{(1)}(\boldsymbol{k}) = \delta_{ij} - 2\frac{k_1 k_j}{k^2}$$
$$M_{ijl}(\boldsymbol{k}) = \tfrac{1}{2}(k_j D_{il}(\boldsymbol{k}) + k_l D_{ij}(\boldsymbol{k})).$$

We have assumed that the condition $\mathrm{Ra}/(k_c L)^4 \ll (\mathrm{Re})^2\mathrm{Pr}_0$ is satisfied (here Ra is the Rayleigh number, Re is the Reynolds turbulence number, $\mathrm{Pr}_0 = \nu_0/\chi_0$ is the Prandtl molecular number, L is the convective layer thickness, and k_c is the characteristic turbulence wave number related to the characteristic scale of energy-carrying vortices of the large-scale turbulence by the formula $k_c = 2\pi/l$). In this case the solutions of equations (5.4.19) and (5.4.20) can be written as

$$\boldsymbol{v}^{T(0)}(\boldsymbol{k};t) \approx \boldsymbol{v}_0^{T(0)}(\boldsymbol{k};t) + \Delta\boldsymbol{v}^{T(0)}(\boldsymbol{k};t)$$
$$\boldsymbol{v}^{T(1)}(\boldsymbol{k};t) \approx \boldsymbol{v}_0^{T(1)}(\boldsymbol{k};t) + \Delta\boldsymbol{v}^{T(1)}(\boldsymbol{k};t).$$

It is assumed that the statistical properties of a non-perturbed field $v_0^{T(0)}$ follow from the expression

$$\left\langle v_{0i}^{T(0)}(\boldsymbol{k};t)v_{0j}^{T(0)}(\boldsymbol{k}';t')\right\rangle = \delta(\boldsymbol{k}+\boldsymbol{k}')\left[D_{ij}(\boldsymbol{k})Q_1(k;t,t') + \mathrm{i}\frac{k_l}{k^2}\varepsilon_{ijl}Q_2(k;t,t')\right]$$

where $Q_1(k;t,t')$ and $Q_2(k;t,t')$ determine the energy and helicity spectra, respectively, of the non-perturbed field. It is also assumed that $\langle\hat{F}(\boldsymbol{k};t,t_1)\rangle = F(k;t,t_1)$.

Using the formulas given above and continuing to stick to the spirit of TSDIA, we obtain from (5.4.15) and (5.4.16)

$$\frac{\partial\langle\boldsymbol{v}\rangle}{\partial t} + \hat{\alpha}\langle\boldsymbol{\Omega}\rangle = -\nabla\langle p\rangle + \hat{\nu}\nabla^2\langle\boldsymbol{v}\rangle + \boldsymbol{g}\beta\langle\theta\rangle \tag{5.4.21}$$

$$\frac{\partial\langle\theta\rangle}{\partial t} = \hat{\chi}\nabla^2\langle\theta\rangle + \boldsymbol{\kappa}\cdot\langle\boldsymbol{v}\rangle \tag{5.4.22}$$

where $\langle\boldsymbol{\Omega}\rangle = \nabla\times\langle\boldsymbol{v}\rangle$; the operators of the renormalized viscosity and heat conduction are given by the relations

$$\hat{\nu}\nabla^2\langle\boldsymbol{v}\rangle = \nu_0\nabla^2\langle\boldsymbol{v}\rangle + \frac{28\pi}{15}\int_{k_c}^{\infty}\mathrm{d}k\cdot k^2\int_{-\infty}^{t}\mathrm{d}t_1 G(k;t,t_1)Q_1(k;t_1,t)\nabla^2\langle\boldsymbol{v}(\boldsymbol{X};t_1)\rangle \tag{5.4.23}$$

$$\hat{\chi}\nabla^2\langle\theta\rangle = \chi_0\nabla^2\langle\theta\rangle + \frac{8\pi}{3}\int_{k_c}^{\infty}\mathrm{d}k\cdot k^2\int_{-\infty}^{t}\mathrm{d}t_1 F(k;t,t_1)Q_1(k;t_1,t)\nabla^2\langle\theta(\boldsymbol{X};t_1)\rangle \tag{5.4.24}$$

and $\hat{\alpha}$ is given by

$$\begin{aligned}\hat{\alpha}\langle\boldsymbol{\Omega}\rangle = -\frac{4\pi}{3}\beta(\boldsymbol{kg})\int_{k_c}^{\infty}\mathrm{d}k\cdot k^2 G(k;t,t_1)\\ \times\left(\int_{-\infty}^{t}\mathrm{d}t_2\int_{-\infty}^{t_2}\mathrm{d}t_3 G(k;t,t_2)F(k;t_2,t_3)Q_2(k;t_1,t_2)\langle\boldsymbol{\Omega}(\boldsymbol{X};t_2)\rangle\right.\\ \left.+\int_{-\infty}^{t_1}\mathrm{d}t_2\int_{-\infty}^{t_2}\mathrm{d}t_3\, G(k;t_1,t_2)F(k;t_2,t_3)Q_2(k;T_3,t)\langle\boldsymbol{\Omega}(\boldsymbol{X};t_3)\rangle\right).\end{aligned} \tag{5.4.25}$$

The quantity $G(k;t,t_1)$ is found from the expression $\left\langle\hat{G}_{ij}(\boldsymbol{k};t,t_1)\right\rangle = D_{ij}(k)G(k;t,t_1)$, where $\left\langle\hat{G}_{ij}(\boldsymbol{k};t,t_1)\right\rangle$ is the Green's function of equation (5.4.20).

We assume that $G(k;t,t_1)$, $F(k;t,t_1)$, $Q_1(k;t,t_1)$ and $Q_2(k;t,t_1)$ in the inertial range of small-scale turbulence can be written as

$$G(k;t,t_1) = H(t-t_1)\exp[-\omega_g(k)(t-t_1)] \tag{5.4.26}$$

$$F(k; t, t_1) = H(t - t_1) \exp[-\omega_f(k)(t - t_1)] \tag{5.4.27}$$

$$Q_1(k; t, t_1) = \sigma(k) \exp[-\omega_1(k)(t - t_1)] \tag{5.4.28}$$

$$Q_2(k; t, t_1) = h(k) \exp[-\omega_2(k)(t - t_1)] \tag{5.4.29}$$

where $H(t - t_1)$ is a step function, and $\sigma(k)$ and $h(k)$ are the spectral densities of turbulent energy and helicity, respectively. Substituting (5.4.26)–(5.4.29) into (5.4.23)–(5.4.25) and integrating over time, we obtain from (5.4.21) and (5.4.22) the set of equations

$$\frac{\partial \langle \boldsymbol{v} \rangle}{\partial t} + \alpha_1 \langle \boldsymbol{\Omega} \rangle = -\nabla \langle p \rangle + \nu_1 \nabla^2 \langle \boldsymbol{v} \rangle + \boldsymbol{g}\beta \langle \theta \rangle \tag{5.4.30}$$

$$\frac{\partial \langle \theta \rangle}{\partial t} = \chi_1 \nabla^2 \langle \theta \rangle + \boldsymbol{\kappa} \cdot \langle \boldsymbol{v} \rangle \tag{5.4.31}$$

where now

$$\nu_1 = \nu_0 + \frac{28\pi}{15} \int_{k_c}^{\infty} \mathrm{d}k\, k^2 \frac{\sigma(k)}{\omega_1(k) + \omega_g(k)}$$

$$\chi_1 = \chi_0 + \frac{8\pi}{3} \int_{k_c}^{\infty} \mathrm{d}k\, k^2 \frac{\sigma(k)}{\omega_1(k) + \omega_f(k)}$$

$$\alpha_1 = \frac{4\pi}{3} \beta(\boldsymbol{kg}) \int_{k_c}^{\infty} \mathrm{d}k\, k^2 \frac{h(k)}{\omega_g(\omega_2 + \omega_f)(\omega_g + \omega_f)}.$$

Let us set $\boldsymbol{g} = (0, 0, -g)$ and $\boldsymbol{\kappa} = (0, 0, \kappa)$. Using dimensionless variables $\tau = t\chi/L^2$, $\boldsymbol{x} = \boldsymbol{X}/L$, $\Theta = \langle \theta \rangle/(\kappa L)$ and $\boldsymbol{V} = \langle \boldsymbol{v} \rangle L/\chi$, and rewriting the velocity as $\boldsymbol{V} = \nabla \times (\psi z) + \nabla \times \nabla \times (\phi \boldsymbol{e})$, where ψ and ϕ are the toroidal and poloidal components of the velocity field, respectively, we can recast the set (5.4.30)–(5.4.31) to the form

$$\left(\frac{\partial}{\partial \tau} - \nabla^2 \right) \nabla^2 \phi - \tilde{\alpha} \nabla^2 \psi + \tilde{\mathrm{R}}\mathrm{a}\Theta = 0 \tag{5.4.32}$$

$$\left(\frac{\partial}{\partial \tau} - \nabla^2 \right) \psi = \tilde{\alpha} \nabla^2 \phi \tag{5.4.33}$$

$$\left(\frac{\mathrm{Pr}\partial}{\partial \tau} - \nabla^2 \right) \theta = -\nabla_\perp^2 \phi \tag{5.4.34}$$

where

$$\tilde{\mathrm{R}}\mathrm{a} = \frac{g\kappa\beta}{\nu\chi} L^4$$

$$\tilde{\alpha} = \frac{4\pi}{3} \tilde{\mathrm{R}}\mathrm{a} \frac{\chi}{L^3} \int_{k_c}^{\infty} \mathrm{d}k \frac{k^2 h(k)}{\omega_g(\omega_2 + \omega_f)(\omega_g + \omega_f)}.$$

The set (5.4.32)–(5.4.34) is considered under the boundary conditions

$$V_z(z = 0, L) = 0$$

$$\frac{\partial V_x}{\partial z}(z = 0, L) = 0$$
$$\frac{\partial V_y}{\partial z}(z = 0, L) = 0$$

and its solution is sought in the form $(\phi, \psi, \Theta) \propto \sin(\pi z)\exp[\mathrm{i}(\boldsymbol{k}_\perp \boldsymbol{r}_\perp + \Gamma t)]$. The non-trivial solution exists if the dispersion relation

$$\widetilde{\mathrm{Ra}}\, k_\perp^2(\Gamma + \pi^2 + k_\perp^2) + \tilde{\alpha}^2(\Gamma \mathrm{Pr} + \pi^2 + k_\perp^2)(\pi^2 + k_\perp^2)^2 \\ - (\Gamma \mathrm{Pr} + \pi^2 + k_\perp^2)(\pi^2 + k_\perp^2)^2(\Gamma + \pi^2 + k_\perp^2) = 0$$

holds. The condition $\Gamma = 0$ yields an expression for the Rayleigh critical number which determines the form of the neutral curve

$$\tilde{\mathrm{Ra}}_{\mathrm{cr}} = \frac{\sqrt{k_\perp^2 + 4\gamma^2(\pi^2 + k_\perp^2)^4} - k_\perp^2}{2\gamma^2(\pi^2 + k_\perp^2)}$$

where

$$\gamma = \frac{4\pi\chi}{3L^3}\int_{k_c}^{\infty} \mathrm{d}k\, \frac{k^2 h(k)}{\omega_g(\omega_2 + \omega_f)(\omega_g + \omega_f)}.$$

As γ increases, the minimal value $\widetilde{\mathrm{Ra}}_{\mathrm{cr,\,min}}$ diminishes while the characteristic horizontal size of the system generated increases.

5.5 Reverse energy cascade in uniform turbulent shear flow

5.5.1 Introduction

We have mentioned earlier that Krause and Rudiger were able to show that helical turbulence cannot amplify perturbations in an incompressible fluid because of the symmetry of the Reynolds stress in the equations for mean fields. If the symmetry of the Reynolds stress is violated, as it is, for example, in compressible or stratified fluids (Moiseev *et al* 1990), the system can amplify large-scale vortical motions that exist at the expense of the energy of small-scale vortices, as a result of their instability.

In this section we will be interested in the evolution of mean perturbations in uniform turbulent shear flow to which we refer, by analogy to the laminar case, as the turbulent Couette flow. It is well known that the plane-parallel laminar Couette flow in unbounded non-viscous fluid is stable with respect to infinitesimal perturbations (Markus and Press 1977); namely, perturbations that grow at an initial stage damp out over long time intervals. This flow is compatible with small-scale, statistically uniform turbulence which now becomes anisotropic. In contrast to the shearless case, it is logical to expect direct energy cascade but with anisotropic viscosity. Helicity properties in the correlator of the ground state may result in energy transfer from smaller to larger scales.

We will now discuss those novel effects that turbulence contributes to the evolution of linear perturbations; in this we follow the work of Moiseev *et al* (1991).

As expected, the non-helical part of the correlator of anisotropic turbulence results in turbulent viscosity, which differs in an anisotropic additional term from the coefficient calculated by Krause and Rudiger (1974). As for the helical component of the correlator, it does not cause mean perturbations growing with time, in contrast to the results of Moiseev *et al* (1990). However, a comparison of the two solutions shows that unlike the laminar flow, in the turbulent Couette flow there is a reverse energy cascade, manifesting itself in much steeper growth of large-scale perturbations at short durations—exponential and even super-exponential, so that even modes that only damp out in the laminar case grow in the turbulent case. Over long times, perturbations again damp out. Three-dimensional (non-planar) structure arises against the background of small-scale turbulence, and its lifetime is determined by the helical component of the correlator. It will be shown below that this structure, stemming from the reverse energy cascade, vanishes in the absence of both helicity and anisotropy in the correlator. Note that both anisotropy and helicity are matters of principle in our formulation of the problem. Anisotropy is a result of a linear velocity shear present in the ground state; in the limit where the anisotropic component of the correlator is ignored, we need to exclude the shear flow as well, and in this situation there can be no reverse cascade even if helicity is non-zero (Moiseev *et al* 1983a).

5.5.2 Formulation of the problem. Basic equations. Homogeneous helical anisotropic flow

In this section we describe the hydrodynamic flow and analyze its stability. We start with the basic Navier–Stokes equations for incompressible flow,

$$\frac{\partial \boldsymbol{V}}{\partial t} + (\boldsymbol{V}\boldsymbol{\nabla})\boldsymbol{V} = -\boldsymbol{\nabla} P + \nu \Delta \boldsymbol{V} + \boldsymbol{F} \tag{5.5.1}$$

$$\boldsymbol{\nabla}\boldsymbol{V} = 0 \tag{5.5.2}$$

where P is pressure and ν is kinematic viscosity.

To describe turbulent flows, we rewrite the hydrodynamic fields as sums of the mean and pulsational components described by the Reynolds equations:

$$\boldsymbol{V} = \bar{\boldsymbol{U}} + \boldsymbol{V}^T \tag{5.5.3}$$

$$P = \bar{P} + P^T \tag{5.5.4}$$

$$\frac{\partial \bar{\boldsymbol{U}}}{\partial t} + (\bar{\boldsymbol{U}}\boldsymbol{\nabla})\bar{\boldsymbol{U}} + \langle(\boldsymbol{V}^T\boldsymbol{\nabla})\boldsymbol{V}^T\rangle = -\boldsymbol{\nabla} P + \nu_0 \Delta \bar{\boldsymbol{U}} \tag{5.5.5}$$

$$\boldsymbol{\nabla}\bar{\boldsymbol{U}} = 0 \tag{5.5.6}$$

$$\frac{\partial \boldsymbol{V}^T}{\partial t} + (\boldsymbol{V}^T\boldsymbol{\nabla})\boldsymbol{V}^T - \langle(\boldsymbol{V}^T\boldsymbol{\nabla})\boldsymbol{V}^T\rangle$$
$$+ (\boldsymbol{V}^T\nabla)U + (U\boldsymbol{\nabla})\boldsymbol{V}^T = -\boldsymbol{\nabla}P^T + \nu_0\Delta\boldsymbol{V}^T + \boldsymbol{f}^T \tag{5.5.7}$$
$$\nabla\boldsymbol{V}^T = 0. \tag{5.5.8}$$

We see from equations (5.5.3)–(5.5.8) that the mean velocity field

$$\bar{U}_i = S\delta_{1i}\delta_{2j}x_j$$

is compatible with the averaged equations (5.5.5) and (5.5.6) (which it satisfies) and with the equations for the pulsational components (5.4.7) and (5.5.8) for uniform turbulence (the equations for the correlators of the random velocity field of such flows are found to have constant coefficients). We thus see that the Reynolds stress for the arbitrary linear profile of mean velocity does not enter the equations for the mean velocity, and that the non-uniform distribution of mean velocity does not affect the correlation properties of turbulence: indeed, the tensor $\partial U_i/\partial x_j$ = const is independent of coordinates. The correlation tensor of turbulence is now anisotropic and we need to fix the expression for it. The general form for the anisotropic correlation tensor can be obtained from phenomenological principles (Batchelor 1953). Note that the turbulence's gyrotropic mode is the simplest form of anisotropy, that is, of absence of mirror symmetry. Another form of anisotropy is the so-called axisymmetric turbulence in which the correlation properties are invariant with respect to the rigid rotations around the axis whose direction is fixed by the unit vector $\boldsymbol{\lambda}$. The general form of the two-point correlation tensor is

$$\hat{Q}_{ij}(\boldsymbol{k}) = A\delta_{ij} + Bk_ik_j + C\varepsilon_{ijk}k_k + D\lambda_i\lambda_j + E\varepsilon_{ijk}\lambda_k + Fk_i\lambda_j + G\lambda_ik_j$$
$$+ H\varepsilon_{ikl}k_kk_j\lambda_l + J\varepsilon_{ikl}k_kk_l\lambda_j + K\varepsilon_{ikl}\lambda_kk_l\lambda_j. \tag{5.5.9}$$

The scalar coefficients A and B are arbitrary functions of $|\boldsymbol{k}|$ and $k_k\lambda_k$. Additional anisotropy, such as the linear shear flow which does not violate the uniformity conditions, forces us to introduce additional directions into the correlation tensors. This increases the number of scalar coefficients; for example, adding another favored direction creates more than ten additional terms in expression (5.5.9).

The problem of establishing the form of the correlation tensor is a fairly cumbersome procedure in the general case. However, one can achieve significant progress by considering turbulent flows with low Reynolds (Re) and Strouhal (Sr) numbers. Taking into account the effect of the averaged motion on turbulence, it is then possible to limit the analysis to the first order of perturbation theory.

With no shear, the Fourier transform of the correlator of uniform, isotropic helical turbulence is

$$Q_{ij}^T(\boldsymbol{k},\omega) = \left(\delta_{ij} - \frac{k_ik_j}{k^2}\right)Q^T(|\boldsymbol{k}|,\omega) + \mathrm{i}\varepsilon_{ijk}k_k\hat{G}(|\boldsymbol{k}|,\omega). \tag{5.5.10}$$

As a result of the effect of shear, the steady-state turbulence state is anisotropic and its correlation tensor has the form

$$
\begin{aligned}
Q_{ij}(\boldsymbol{k},t) &= Q_{ij}^{T}(\boldsymbol{k},t) \\
&+ SQ'(\boldsymbol{k},S,t)(\delta_{im}\,\delta_{jn}+\delta_{in}\,\delta_{jm})\left(\delta_{1m}-\frac{k_1k_m}{k^2}\right)\left(\delta_{2n}-\frac{k_2k_n}{k^2}\right) \\
&+ \mathrm{i}SG^{1}(\boldsymbol{k},S,t)(\delta_{im}\delta_{jn}-\delta_{in}\delta_{jm})\left(\delta_{1n}-\frac{k_1k_n}{n^2}\right)\varepsilon_{m2f}k_f+\ldots.
\end{aligned}
\tag{5.5.11}
$$

The correlation tensor of the uniform turbulent shear flow can be presented as a sum of isotropic and anisotropic homogeneous parts. Note that the gyrotropic part of the correlator also gains an additional anisotropic term.

5.5.3 Closing the equations for perturbations of turbulent shear flow in the second-order correlation approximation

When taking up the matter of linear shear of mean velocity and uniform turbulence as the ground state, we question its stability with respect to the perturbations of the mean velocity field.

Let a small non-uniform perturbation of mean velocity $\boldsymbol{V}$ arise at moment $t=0$. As a result of large-scale velocity perturbations, a non-uniform component of turbulence $\boldsymbol{V}'$ arises. In the second-order correlation approximation (Krause and Rudiger 1974, Krause and Radler 1980), min Re, Sr $\ll 1$ we obtain for $\boldsymbol{V}$ and $\boldsymbol{V}'$

$$
\frac{\partial V_i}{\partial t}+\hat{S}_{km}x_m\frac{\partial V_i}{\partial x}+\hat{S}_{ik}V_k+\left(\left\langle V_k'\frac{\partial V_i^T}{\partial x_k}\right\rangle+\left\langle V_k^T\frac{\partial V_i'}{\partial x_k}\right\rangle\right)=-\frac{\partial P}{\partial x_i}+\nu_0\Delta V_i
\tag{5.5.12}
$$

$$
\frac{\partial V_i}{\partial x_i}=0
\tag{5.5.13}
$$

$$
\frac{\partial V_i}{\partial t}+\hat{S}_{km}x_m\frac{\partial V_i'}{\partial x_k}+\hat{S}_{ik}V_k'+\left(V_k'\frac{\partial V_i^T}{\partial x_k}+V_k^T\frac{\partial V_i'}{\partial x_k}\right)=-\frac{\partial P}{\partial x_i}+\nu_0\Delta V_i'
\tag{5.5.14}
$$

$$
\frac{\partial V_i'}{\partial x_i}=0.
\tag{5.5.15}
$$

To close the set (5.5.12), (5.5.13) we need to calculate the non-uniform correlation tensor $\langle V_m' V_n^T\rangle$. The solution of the set (5.5.14), (5.5.15) is found more conveniently in the coordinate system co-moving with the mean flow. The transition to the appropriate frame of reference is performed as follows (Krause and Radler 1980). We introduce a matrix $\gamma_{ij}(t)$ that satisfies equation

$$
\frac{\partial \gamma_{ij}}{\partial t}=\hat{S}_{ik}\gamma_{kj}.
\tag{5.5.16}
$$

Here $\hat{S}_{ik}$ is the matrix determined by the gradients of the mean flow:

$$\hat{S}_{ik} = \frac{\partial \bar{U}_i}{\partial x_i}. \tag{5.5.17}$$

Equation (5.5.16) has the following solution for small t:

$$\gamma_{ij}(t) = \delta_{ij} + \hat{S}_{ij} t. \tag{5.5.18}$$

Note that in the case of linear shear the expression (5.5.18) is the exact solution of equation (5.5.16). We introduce a coordinate $\boldsymbol{y}$ that is related to the fixed coordinate $\boldsymbol{x}$ by the equations

$$x_i = \gamma_{ij}(t) y_i \tag{5.5.19}$$

$$y_i = \gamma_{ij}(-t) x_j. \tag{5.5.20}$$

We denote the velocity perturbation in the co-moving frame of reference by $\boldsymbol{V}(\boldsymbol{y}, t)$ and that in the fixed frame of reference by $\boldsymbol{U}(\boldsymbol{x}, t)$ (likewise, $\boldsymbol{f}(\boldsymbol{y}, t)$ and $\boldsymbol{f}'(\boldsymbol{x}, t)$ for the forces). The relations connecting them are

$$U_i = \gamma_{ij}(t) V_j \tag{5.5.21}$$

$$V_i = \gamma_{ij}(-t) U_j \tag{5.5.22}$$

$$f_i' = \gamma_{ij}(t) f_j \tag{5.5.23}$$

$$f_i = \gamma_{ij}(-t) f_j'. \tag{5.5.24}$$

It is not difficult to derive the appropriate equation for the velocity perturbation in the co-moving coordinate system. Indeed,

$$\begin{aligned} \frac{\partial V_i(\boldsymbol{y}, t)}{\partial t} &= \frac{\partial \gamma_{ij}(-t)}{\partial t} U_j(\boldsymbol{x}, t) + \gamma_{ij}(-t) \frac{\partial U_j(\boldsymbol{x}, t)}{\partial t} \\ &= \gamma_{ij}(-t) \left[\frac{\partial U_j(\boldsymbol{x}, t)}{\partial t} + \hat{S}_{lk} x_k \frac{\partial U_i(\boldsymbol{x}, t)}{\partial x_i} - \hat{S}_{il} U_l(\boldsymbol{x}, t) \right]. \end{aligned} \tag{5.5.25}$$

Finally, the equation for the velocity perturbations in the co-moving coordinates $(\boldsymbol{y}, t)$ takes the form

$$\begin{aligned} &\frac{\partial V_i}{\partial t} + 2\hat{S}_{ik} V_k - \nu_0 \gamma_{mr}(-t) \gamma_{nr}(-t) \frac{\partial^2 V_i}{\partial y_m \, \partial y_n} \\ &\qquad = -\gamma_{ir}(-t) \gamma_{kr}(-t) \frac{\partial P}{\partial y_k} + f_i(\boldsymbol{y}, t) \end{aligned} \tag{5.5.26}$$

$$\frac{\partial V_k}{\partial y_k} = 0. \tag{5.5.27}$$

By performing the spatial Fourier transform and eliminating pressure we can write

$$\frac{\partial \hat{V}_i(\boldsymbol{k},t)}{\partial t} + \nu_0\gamma_{mr}(-t)\gamma_{nr}(-t)k_m k_n \hat{V}_i(\boldsymbol{k},t)$$
$$+\left[\delta_{ip} - \frac{\gamma_{im}(-t)\gamma_{mn}(-t)k_n k_p}{\gamma_{mr}(-t)\gamma_{nr}(-t)k_m k_n}\right]\left[2S\delta_{p1}\hat{V}_2(\boldsymbol{k},t) - f_p(\boldsymbol{k},t)\right] = 0. \tag{5.5.28}$$

$$k_i V_i = 0. \tag{5.5.29}$$

Since the final goal of the calculations is to obtain the Reynolds stress $\langle V'_m V^T_n\rangle$, we need to remark that within our approximation (small Strouhal or Reynolds numbers), it is non-zero only on the time scale $t - t' \ll \tau_{cor}$ and therefore, to solve the equation for $\boldsymbol{V}'$ it is sufficient to ignore the dependence of the coefficients of equations (5.5.28) and (5.5.29) on time, that is, to retain only the first powers of the expansion into the Taylor series in $S\tau_{cor} \ll 1$ when calculating the correlator (this inequality is implied by the two-scale nature of our problem). This approximation is similar to the method used to calculate the effect of various perturbations on the turbulent fluxes (the so-called Rapid Distortion Turbulence (Townsend 1976, Hunt and Carruthes 1990) or first approximation in the mean velocity field (Krause and Radler 1980)).

Having now carried out the Fourier transform in time, we can write the equation for $\boldsymbol{V}$ in the $(\boldsymbol{k},\omega)$ representation:

$$\left[-\mathrm{i}\omega + \nu_0 k^2\right]\hat{V}'_i(\boldsymbol{k},\omega) + 2S\left[\delta_{i1} - \frac{k_i k_1}{k^2}\right]\hat{V}'_2(\boldsymbol{k},\omega)$$
$$= -\mathrm{i}\left[\delta_{ip} - \frac{k_i k_p}{k^2}\right]\left[\delta_{km}\delta_{pn} + \delta_{kn}\delta_{pm}\right]k_k \int_{-\infty}^{+\infty}\int_{-\infty}^{+\infty} \mathrm{d}\boldsymbol{k}'\,\mathrm{d}\omega'$$
$$\times \hat{V}^T_m(\boldsymbol{k}-\boldsymbol{k}',\omega'-\omega)\hat{V}'_n(\boldsymbol{k}',\omega') \tag{5.5.30}$$

$$k_i \hat{V}'_i = 0. \tag{5.5.31}$$

Note that in this approximation $\boldsymbol{V}^T$ and $\boldsymbol{V}$ that enter (5.5.30) are considered in the fixed coordinate system.

The solution of (5.5.30), (5.5.31) is

$$\hat{V}'_i(\boldsymbol{k},\omega) = -\mathrm{i}R_{sp}(\boldsymbol{k},\omega)\left[\delta_{ip} - \frac{k_i k_p}{k^2}\right]\left[\delta_{km}\delta_{pn} + \delta_{kn}\delta_{pm}\right]k_k$$
$$\times \int_{-\infty}^{+\infty}\int_{-\infty}^{+\infty} \mathrm{d}\boldsymbol{k}'\,\mathrm{d}\omega'\,\hat{V}^T_m(\boldsymbol{k}-\boldsymbol{k}',\omega-\omega')\hat{V}'_n(k',\omega') \tag{5.5.32}$$

where

$$R_{sp}(\boldsymbol{k}',\omega') = \frac{1}{(\nu_0 k'^2 - \mathrm{i}\omega')}\left[\delta_{sp} - 2S\frac{\delta_{s1} - (k'_s k'_1/k'^2)}{(\nu_0 k'^2 - 2S(k'_1 k'_2/k'^2) - \mathrm{i}\omega')}\delta_{2p}\right]. \tag{5.5.33}$$

The Reynolds stress

$$\left(\left\langle V_k' \frac{\partial V_i^T}{\partial x_k}\right\rangle + \left\langle V_k^T \frac{\partial V_i'}{\partial x_k}\right\rangle\right)$$

in the Fourier transform representation has the form

$$\langle \hat{T}(\boldsymbol{k}, \omega)\rangle_i = -k_r \hat{V}_n(\boldsymbol{k}, \omega)(\delta_{rs}\delta_{it} + \delta_{rt}\delta_{is})(\delta_{fm}\delta_{pn} + \delta_{fn}\delta_{pm})$$
$$\times \int_{-\infty}^{+\infty}\int_{-\infty}^{+\infty} \mathrm{d}\boldsymbol{k}'\,\mathrm{d}\omega' R_{sp}(\boldsymbol{k}', \omega') + k_f' \hat{Q}_{tm}(\boldsymbol{k} - \boldsymbol{k}', \omega - \omega'). \tag{5.5.34}$$

Here we have omitted the terms of order greater than first order in $S\tau_{\mathrm{cor}}$.

The wave vector $\boldsymbol{k}'$ and frequency ω' in expression (5.5.34) refer to small-scale turbulent pulsations, and the wave vector $\boldsymbol{k}$ and frequency ω refer to large-scale ones. By virtue of the two-scale hypothesis, we can set

$$|\boldsymbol{k}'| \gg |\boldsymbol{k}| \qquad |\omega'| \gg |\omega|. \tag{5.5.35}$$

This allows us to use the following approximation:

$$\hat{Q}_{tm}(\boldsymbol{k} - \boldsymbol{k}', \omega - \omega') \cong \hat{Q}_{tm}(-\boldsymbol{k}', \omega') + \frac{\partial \hat{Q}_{tm}(-\boldsymbol{k}_1' - \omega')}{\partial k_l'} k_l. \tag{5.5.36}$$

Note that in calculations in the (x, t) representation this approximation corresponds to the replacement of expressions of the type of

$$\int_{-\infty}^{+\infty}\int_0^t \mathrm{d}\boldsymbol{x}'\,\mathrm{d}t'\, Q_{sp}(\boldsymbol{x}', t')\,\frac{\partial V_n(\boldsymbol{x} - \boldsymbol{x}', t - t')}{\partial (x - x')_r}$$

in the final result by

$$\frac{\partial V(\boldsymbol{x}, t)}{\partial x_r}\int_{-\infty}^{+\infty}\int_0^t \mathrm{d}\boldsymbol{x}'\,\mathrm{d}t'\, Q_{sp}(\boldsymbol{x}', t')$$

which is also an application of the two-scale hypothesis—relatively slow and smooth variation of large-scale components in comparison with the pulsating small-scale quantities.

In expression (5.5.34) we need to carry out averaging over space. We will use the following obvious relations: $\mathrm{d}\boldsymbol{k} = k^2\,\mathrm{d}k\,\mathrm{d}\Omega$ (Ω is the solid angle and k the modulus of the wave vector),

$$\int k_i k_j\,\mathrm{d}\Omega = \frac{4\pi}{3}k^2\delta_{ij} \tag{5.5.37}$$

$$\int k_i k_j k_l k_m\,\mathrm{d}\Omega = \frac{4\pi}{15}k^4\left[\delta_{ij}\delta_{lm} + \delta_{il}\delta_{jm} + \delta_{im}\delta_{ij}\right]. \tag{5.5.38}$$

After averaging over space, we arrive at the following relation:

$$\begin{aligned}\langle \hat{T}(\boldsymbol{k}.\omega)\rangle_i &= -\mathrm{i}\alpha\,[\delta_{it}\delta_{1r} + \delta_{rt}\delta_{1i}]\,\varepsilon_{t2n}k_r\hat{V}_n(\boldsymbol{k},\omega) - \nu_T k^2 \hat{V}_i(\boldsymbol{k},\omega)\\ &\quad + \nu_T^a\,[2.28\delta_{i1}\delta_{mr}\delta_{2n} - 0.53\delta_{i2}\delta_{mr}\delta_{1n} + 1.03\delta_{ir}\delta_{1m}\delta_{2n}\\ &\quad -1.20\delta_{ir}\delta_{2m}\delta_{1n} + 0.17\delta_{1m}\delta_{2r}\delta_{ni}]\,k_m k_r \hat{V}_n(\boldsymbol{k},\omega).\end{aligned} \tag{5.5.39}$$

Here

$$\alpha = \frac{32\pi S}{15}\int_{-\infty}^{+\infty}\int_{-\infty}^{+\infty} \mathrm{d}q\,\mathrm{d}\omega\,\frac{q^4 G(q,\omega)}{(\nu_0 q^2 - i\omega)^2} \tag{5.5.40}$$

$$\nu_T = \frac{24\pi}{5}\int_{-\infty}^{+\infty}\int_{-\infty}^{+\infty} \mathrm{d}q\,\mathrm{d}\omega\,\frac{q^2 \hat{Q}(q,\omega)}{(\nu_0 q^2 - i\omega)} \tag{5.5.41}$$

$$\nu_T^a = \frac{(4/3)\pi S}{3}\int_{-\infty}^{+\infty}\int_{-\infty}^{+\infty} \mathrm{d}q\,\mathrm{d}\omega\,\frac{q^2 \hat{Q}(q,\omega)}{(\nu_0 q^2 - i\omega)^2}. \tag{5.5.42}$$

We see that taking into account the interaction between large-scale perturbations with small-scale gyrotropic turbulence and shear flow results, in addition to causing anisotropic turbulent viscosity σ, also produces additional terms proportional to the first derivatives of perturbations. Note that the anisotropic gyrotropic component of the correlator has not led to terms of first order in $\boldsymbol{k}$.

5.5.4 Interval instability of turbulent Couette flow

Generation of coherent structures.

We will investigate the stability of large-scale perturbations of turbulent shear flow taking into account the calculated Reynolds stress. The anisotropic component of turbulence viscosity can be treated as negligibly small. The equation for the perturbation thus has the form

$$\begin{aligned}&\frac{\partial V_i}{\partial t} + Sx_2\frac{\partial V_i}{\partial x_i} + S\delta_{i1}V_2 + \alpha S(\delta_{it}\delta_{1r} + \delta_{i1}\delta_{tr})\varepsilon_{t2n}\frac{\partial V_n}{\partial x_r}\\ &\quad = -\frac{\partial P}{\partial x_i} + \nu^*\Delta V_i\end{aligned} \tag{5.5.43}$$

$$\frac{\partial V_i}{\partial x_i} = 0. \tag{5.5.44}$$

The solution of set (5.5.43), (5.5.44) is obtained easier in the co-moving coordinate system; the transition to it was shown in the preceding section. We denote the velocity in the co-moving frame of reference by $\boldsymbol{V}$. The equation for $\boldsymbol{V}$ in the Fourier transform representation is

$$\frac{\partial \hat{V}_i(\boldsymbol{k},t)}{\partial t} + \nu_0 \gamma_{mr}(-t)\,\gamma_{nr}(-t) k_m k_n \hat{V}_i(\boldsymbol{k},t)$$
$$+ \left[\delta_{1p} - \frac{\gamma_{im}(-t)\gamma_{mn}(-t)k_n k_p}{\gamma_{mr}(-t)\gamma_{nr}(-t)k_m k_n}\right]\Big[2S\delta_{p1}\hat{V}_2(\boldsymbol{k},t)$$
$$+ \mathrm{i}\alpha S(\delta_{pt}\delta_{1f} + \delta_{p1}\delta_{tf})\varepsilon_{t2l}\gamma_{fj}(-t)k_j \hat{V}_l\Big] = 0 \tag{5.5.45}$$

$$k_i \hat{V}_i = 0. \tag{5.5.46}$$

Pressure has been eliminated from equations (5.5.45), (5.5.46).

With zero helicity ($\alpha = 0$), the stability problem has the following solution (see also Moiseev *et al* 1991):

$$\hat{V}_j(t) = G_{lj}(t)\hat{V}_l(0) \tag{5.5.47}$$

$$\hat{G}(t) = \begin{bmatrix} 1 & -2\int_0^\tau \left[1 - \frac{k_1^2}{k^2} - \frac{k_1 k_2}{k^2}\theta\right] K^{-1}(\theta)\,\mathrm{d}\theta & 0 \\ 0 & \left[1 - 2\frac{k_1 k_2}{k^2}\tau + \frac{k_1^2}{k^2}\tau^2\right] & 0 \\ 1 & 2\frac{k_1 k_3}{k^2}\int_0^\tau K^{-1}(\theta)\,\mathrm{d}\theta & 1 \end{bmatrix} K(\tau). \tag{5.5.48}$$

Here

$$\tau = S\cdot t$$
$$K(\tau) = \exp\left[-\frac{\nu_T}{S}k^2\tau \cdot \left[1 - \frac{k_1 k_2}{k^2}\tau + \frac{k_1^2}{k^2}\tau^2\right]\right]. \tag{5.5.49}$$

The expressions (5.5.47), (5.5.48) give an answer to the question about the linear stability of planar unbounded Couette flow (Markus and Press 1977). The perturbations grow at the initial stage $St \leqslant |k_1/k_2|$ but then decay, the decay being steeper owing to the shear ($\approx \exp(-\nu_T k_1^2 S^2 t^3/3)$). The equations for the evolution of perturbations can be reduced to a set of two coupled equations for $\hat{\boldsymbol{V}}$ and rot $\hat{\boldsymbol{V}}$:

$$\frac{\partial \hat{V}_2}{\partial \tau} - 2\left[k_1^* k_2^* - k_1^{*2}\tau\right]\hat{V}_2 + \nu_* k^{**2}\hat{V}_2$$
$$+ 2\alpha\left[k_1^* - k_2^*\tau\right]\left[k_2^* - k_1^*\tau\right]\hat{W}_2 = 0 \tag{5.5.50}$$

$$\frac{\partial \hat{W}_2}{\partial \tau} - \mathrm{i}\alpha k_3\left[1 + 2\tau\left[k_1^* - k_2^*\tau\right]\left[k_2^* - k_1^*\tau\right]\right]\hat{W}_2 + \nu_* k^{**2}\hat{W}_2$$
$$+ \left[\alpha k_2\left[k_1 - 2k_2\tau\right] + 2\mathrm{i}k_3\left[k^{*2} - k_1^* k_2^*\tau\right]\right]\hat{V}_2 = 0. \tag{5.5.51}$$

We have introduced above the following notation:

$$k_1^* = \frac{k_1}{k^{**}}$$

$$k^{**} = \left[k^2 - 2k_1k_2\tau + k_1^2\tau^2\right]^{1/2} \quad (5.5.52)$$

$$k^{*2} = k_n^*k_n^* \quad (5.5.53)$$

$$\hat{W}_2 = \mathrm{i}(k_3\hat{V}_1 - k_1\hat{V}_3). \quad (5.5.54)$$

The replacement

$$\hat{V}_2 = \varphi(\tau)K(\tau) \qquad \hat{W}_2 = \psi(\tau)K(\tau) \quad (5.5.55)$$

removes from (5.5.50), (5.5.51) the dissipative terms, otherwise the analysis of this set in the general case would be too unwieldy.

We choose two limiting cases for analysis

(A) $k_3 = 0$; $k^2 = k_1^2 + k_2^2$. In this case the set of equations for φ and ψ is found to be

$$\frac{\partial\varphi}{\partial\tau} - 2\left(k_1^*k_2^* - k_1^{*2}\tau\right)\varphi + 2\alpha\left[k_1^* - k_2^*\tau\right]\left[k_2^* - k_1^*\tau\right]\psi = 0 \quad (5.5.56)$$

$$\frac{\partial\psi}{\partial\tau} + \alpha k_2\left[k_1 - 2k_2\tau\right]\varphi = 0. \quad (5.5.57)$$

The set (5.5.56), (5.5.57) reduces to a single equation of second order for the function $\Phi = k^{**2}\varphi$,

$$\frac{\partial^2\Phi}{\partial\tau^2} + \frac{\left[k_1^2 + k_2^2 - 2k_1k_2\tau\right]}{[k_1 - k_2\tau][k_2 - k_1\tau]}\frac{\partial\Phi}{\partial t} - 2\alpha^2k_2\frac{[k_1 - 2k_2\tau][k_1 - k_2\tau][k_2 - k_1\tau]}{k_1^2 + (k_2 - k_1\tau)^2}\Phi = 0. \quad (5.5.58)$$

We introduce a dimensionless parameter $\beta = k_1/k_2$; $\beta \ll 1$ for most perturbations considered for the shear flow. Denoting $\alpha^* = \alpha k_2$, we rewrite equation (5.5.58) as

$$\frac{\partial^2\Phi}{\partial\tau^2} + \frac{[1 + \beta^2 - 2\beta\tau]}{[\beta - \tau][1 - \beta\tau]}\frac{\partial\Phi}{\partial\tau} - 2\alpha^{*2}\frac{[\beta - 2\tau][\beta - \tau][1 - \beta\tau]}{\beta^2 + [1 - \beta\tau]^2}\Phi = 0. \quad (5.5.59)$$

The evolution of perturbations in a shear flow occurs over time of the order of $\beta\tau \approx 1$; we see power-growth followed by rapid exponential decay. A characteristic feature of equation (5.5.59) is a very weak dependence of the behavior of the perturbation on β over times $\beta \ll \tau \ll \beta^{-1}$ ($\beta \ll 1$); this behavior can be found from the equation

$$\frac{\partial^2\Phi}{\partial t^2} - \frac{1}{\tau}\frac{\partial\Phi}{\partial\tau} - 4\alpha^{*2}\tau^2\Phi = 0. \quad (5.5.60)$$

The solutions of equation (5.5.60) are the Bessel functions of imaginary argument:

$$\Phi(\tau) = \tau Z_{1/2}(\mathrm{i}\alpha^*\tau^2). \quad (5.5.61)$$

Therefore, if $\beta \ll \tau \ll \beta^{-1}$ ($\beta \ll 1$), we have

$$\hat{V}_2(\boldsymbol{k}, \tau) \approx A(k_2, \beta)\tau Z_{1/2}(\mathrm{i}\alpha k_2 \tau^2) \exp(-\nu_T k_2^2 \tau / S) + O(\beta k_2, \beta\tau) \tag{5.5.62}$$

where $A(k_2, \beta) = \text{const}\, k_2^{-2}$.

Taking into account that at large τ

$$Z_{1/2}(\mathrm{i}\alpha k_2 \tau^2) \approx \tau^{-1/2} \exp(\alpha k_2 \tau^2)$$

(Nikiforov and Uvarov 1984), we find

$$\hat{V}_2 \approx A(k_2, \beta)\tau^{1/2} \exp(\alpha k_2 \tau^2 - \nu_T k_2^2 \tau / S). \tag{5.5.63}$$

The perturbation undergoes a faster-than-exponential growth for a considerable period of time and, before dissipative mechanisms come into play, its amplitude increases enough to take the problem outside the applicability realm of the linear theory. If $\beta\tau = \theta \gg 1$, equation (5.5.59) also becomes of Bessel type:

$$\frac{\partial^2 \Phi}{\partial \theta^2} - \frac{2}{\theta}\frac{\partial \Phi}{\partial \theta} + 4\frac{\alpha^{*2}}{\beta^2}\theta^2 \Phi = 0. \tag{5.5.64}$$

Its solutions are the Bessel functions of real argument:

$$\Phi(\theta) = \theta^{3/2} Z_1 \left[\frac{4\alpha^*}{3\beta} \tau^{3/2} \right]. \tag{5.5.65}$$

Using the asymptotic representation for the Bessel functions (Nikiforov and Uvarov 1984) we find that if $\beta t \gg 1$, perturbations decay as

$$\hat{V}_2 \approx B(k_1, k_2, \alpha)\tau^{-5/4} \exp(-\nu_T k_1^2 \tau^3 / S) \cos\left[\frac{4}{3}\alpha (k_1 k_2)^{1/2} \tau^{3/2} + \frac{3\pi}{4} \right]. \tag{5.5.66}$$

Therefore, flow perturbations for which

$$\beta = \frac{k_1}{k_2} \ll 1$$

grow over time $\tau \ll \beta^{-1}$, and the steepest growth rate occurs for perturbations with $k_1 \approx (\alpha S / 2\nu_T)$; they increase by a factor $\approx \exp[(\alpha^2 S / 4\nu_T)\beta^{-3}]$. Growth ceases over a time $\tau \approx \beta^{-1}$ and perturbations decay over time $\tau \gg \beta^{-1}$. The time evolution of perturbations in a shear flow is of the interval instability type. However, the super-exponential takes us beyond the region of validity of the linear instability theory.

A numerical analysis of equations (5.5.56), (5.5.57) confirmed the increase of amplitude at the growth stage by up to two orders of magnitude in comparison with the evolution when there is no helicity (Marcus and Press 1977).

(B) $k_1 = 0$; $k^2 = k_2^2 + k_3^2$. In this case the set of equations for φ and ψ can be written as

$$\frac{\partial\varphi}{\partial\tau} - 2\alpha k_2^{*2}\tau\psi = 0 \tag{5.5.67}$$

$$\frac{\partial\psi}{\partial\tau} - \mathrm{i}\alpha k_3\left[1 - 2\frac{k_2^2}{k^2}\tau^2\right]\psi + 2\left[\mathrm{i}k_3 - \alpha k_2^2\tau\right]\varphi = 0. \tag{5.5.68}$$

Transforming equations (5.5.67) and (5.5.68) to a single equation and performing a replacement of a variable

$$\theta = \mathrm{i}\tau$$

we find an equation for φ:

$$\frac{\partial^2\varphi}{\partial\theta^2} - \left[\frac{1}{\theta} + \alpha k_3\left[1 + 2\frac{k_3}{k^2}\theta^2\right]\right]\frac{\partial\varphi}{\partial\theta} - 4\alpha\frac{k_3}{k^2}\left[\alpha k_2^2\theta^2 + k_3\right] = 0. \tag{5.5.69}$$

The solutions of equation (5.5.69) are:
for $\theta \ll 1$

$$\varphi \cong \mathrm{i}\tau Z_{2/3}\left(-\frac{4}{3}\left(\frac{\mathrm{i}}{2}\alpha k_3\right)^{1/2}\tau^3\right) \tag{5.5.70}$$

for $\theta \gg 1$

$$\varphi \cong \mathrm{i}\tau \exp\left(\frac{\mathrm{i}}{3}\alpha k_3\frac{k_2^2}{k^2}\tau^3\right) Z_{1/6}\left(-\frac{\alpha}{3}k\frac{k_2^2}{2}\tau^3\right). \tag{5.5.71}$$

This means that over short times τ perturbations increase in proportion to τ^2 and oscillate, but decay over longer times. Note that in this case only decaying modes existed when helicity was absent.

The analysis given above showed the dominant influence of uniform shear turbulence on the evolution of the averaged large-scale fields. The interaction between the mean perturbations and a given small-scale flow produces non-uniform additional terms that determine Reynolds stress in averaged equations. Note also that helicity stimulates the growth of those components that normally only decayed. The super-exponential growth of three-dimensional perturbations at the early stages rapidly moves us out of the range of applicability of the linear approximation. To give the ultimate answer on the existence of long-lived coherent structures, we need to take into account nonlinear terms. As for the helicity properties of small-scale turbulence, they definitely accelerate the growth of perturbations via energy transfer from small scales to large scales. This reverse energy cascade now determines at what moment the nonlinear stage is achieved.

To conclude this subsection, we wish to emphasize that correct formulation of the problem of instability of ground state made it possible to demonstrate the instability of the Couette flow under helicity turbulence.

5.6 Helicity as a possible physical mechanism for the amplification of typhoon-type vortical perturbations in the atmosphere and of the modulation of rotation frequency of the Earth

5.6.1 Introductory remarks

The preceding sections of this chapter showed that large-scale structures may be generated in a medium with small-scale helical turbulence. The main feature of this theory, which distinguishes it from other models of large-scale structure generation, is the dissipation feedback between the solenoidal velocity components; this happens owing to the helicity of the small-scale field. This is the factor that leads to a new type of instability. Note also that on the whole, helical vortices have a stronger tendency to merge than an ensemble of ordinary vortices. It must also be emphasized that at both large and small Reynolds numbers (see above) the type of both feedback and instability stemming from the helicity of small-scale motion remain the same regardless of which approximation is used.

Even though these arguments boost confidence in the correctness of our analysis, we cannot consider the study complete, even for the linear stage of the long-wavelength instability that is of interest to us here. For example, there remains a danger of secular terms appearing in some cases. Furthermore, the analysis of higher-order moments has not been completed. On the other hand, taking into account even the fourth moment in the so-called τ-approximation shows that the instability of second moments arises due to the helical instability of second moments in situations where this is impossible in the second correlation approximation (see Belian *et al* 1996).

The theory is thus vigorously progressing and it is all the more important to evaluate to what extent it is supported by experimental results. It is of special interest to turn to geophysical applications, in view of their practical significance. We will apply the concept of helicity to analyzing the generation of typhoons and the mechanism of modulation of the number of strong earthquakes and the frequency of rotation of the Earth. The details will be found in the subsections that follow but we need to note here that the events that we are to consider are certainly not the only possible objects for geophysical applications. Thus the evolution of a cyclone in mid latitudes can also be given a plausible explanation by appealing to the helicity of small-scale turbulence, assuming it was caused by the instability of shear flow.[3] Finally, not only geophysical but numerous other astrophysical, heliophysical, comet and planetary phenomena are also governed by helicone processes. As an example, we can point to the evolution of Jupiter's atmosphere after the collision with the Shoemaker–Levy 9 comet. Fortov *et al* (1996) were able to show that the typhoon (helical) mechanism was responsible for the subsequent scenario of processes in the Jovian atmosphere.

[3] This aspect was first considered by Moiseev, Sagdeev and Chkhetiani (unpublished).

5.6.2 Qualitative analysis of the role played by helicity in typhoon generation

A well developed tropical typhoon (hurricane) is an intense vortex in which the main velocity component lies in the horizontal plane. A weak—but critically important for the system as a whole—vertical circulation is superposed on the high-power horizontal circulation. Many years of studies of tropical cyclones have accumulated a rich body of observations whose generalization allows one to build numerical models that describe the structure of a fully developed typhoon.

It still remained unclear until very recently what physical processes occur in the initial stages of the evolution of a tropical cyclone, how a 'coupled' system (toroidal plus poloidal velocity fields) appears in a natural way, and whether it is possible to assure amplification of the initial perturbation through this feedback. In numerical simulations, these questions are circumvented by choosing the initial conditions that typically contain the main features of a mature typhoon (see, for example, Riel 1976). The analytical models outlined above have the drawback of using unsubstantiated hypotheses on the structure of convective flows (see, for example, Khain and Sutyrin 1983). In other words, the existing models of the initial stages of typhoon evolution are phenomenological and cannot pretend to explain the mechanism of formation of tropical cyclones (TC) from first principles. The complexity of the models manifests itself not only in a heavy admixture of phenomenology but also in 'pumping up' the number of physical factors required for the growth of a TC. We find it, for example, in a largely attractive theory known as a conditional second-order instability (see, for example, Ooyama 1964).

According to this theory, a TC requires for its growth a depression-type structure (initial stage of TC), unstable flow in the surface layer of the ocean, and humid convection. In addition, the vertical flows due to humid convection are parametrized (that is, certain phenomenological assumptions are made about their structure). The question is: can we minimize the constraints on the model of TC generation? We fully realize that the process of inception and growth of a tropical cyclone in the real atmosphere is much more complicated than any theoretical construct; nevertheless, we believe that a successful 'minimum model' clarifies the physical side of the phenomena. From this standpoint, it is first of all necessary to single out one essential feature of tropical cyclones. This crucial feature is the helical nature of the velocity field in them, that is, the linkage of the streamlines, which can be written as the following property of the velocity field $\langle \boldsymbol{v} \rangle$: $\int \langle \boldsymbol{v} \rangle \, \mathrm{rot} \langle \boldsymbol{v} \rangle \, \mathrm{d}^3 r \neq 0$ (for information on the topological properties of helical flows see, for example, Moffat 1978). Note that the flow helicity $\int \langle \boldsymbol{v} \rangle \, \mathrm{rot} \langle \boldsymbol{v} \rangle \, \mathrm{d}^3 r$ is a non-viscous integral of motion and that the generation of motions with linked streamlines requires special explanation. For this reason a theory of formation of a tropical cyclone must primarily explain this topological property of the velocity field.

Moiseev *et al* (1983a) suggested treating the formation of a TC as a secondary unstable process that evolves against the background of small-scale helical turbulence in the atmosphere. The reader familiar with the preceding sections of this chapter will easily be able to check that in this case the linkage arises in the large-scale velocity field quite naturally, since the exponentially growing solution is indeed helical. We thus see that the topological non-triviality of the streamlines in tropical cyclones is related in principle to the non-zero mean helicity of the small-scale velocity field. A spectacular feature is the fact that owing to Earth's rotation, the small-scale turbulent ensemble possesses non-zero mean helicity (see section 5.2).

As a rule, tropical cyclones are born in regions with intense convective motions in the atmosphere. Typhoons do not arise at the equator or at mid latitudes: the band of TC inception is bounded by the latitudes 5° and 25°. An ensemble of convective cells is therefore an acceptable candidate that provides a feedback between toroidal and poloidal motion in a TC.

Still within our qualitative analysis, we take the equation for the mean vorticity of a large-scale velocity field $\langle \boldsymbol{\Omega} \rangle = \mathrm{rot}\,\langle \boldsymbol{v} \rangle$ in its simplest form,

$$\frac{\partial \langle \boldsymbol{\Omega} \rangle}{\partial t} = \mathrm{rot}\,\langle \alpha \boldsymbol{\Omega} \rangle + \nu \Delta \langle \boldsymbol{\Omega} \rangle \tag{5.6.1}$$

where ν is the turbulent viscosity coefficient, $\alpha \approx (-\tau/3)\langle \boldsymbol{v}\ \mathrm{rot}\ \boldsymbol{v} \rangle$ is a parameter, and τ is the correlation time.

It will be clear from what follows later that we make an attempt to take into account explicitly the role of convective instability; this brings us closer to the actual tropical situation. In this section, however, we pay maximum attention to the relative simplicity of the 'minimum model' (requiring only the small-scale helical turbulence) and also its adequate rigorousness (the form of Reynolds stress is obtained from first principles, not phenomenologically).

To evaluate the parameters of the velocity field generated, we consider a mathematically simple case that allows analytical solution. We will analyze an asymptotic solution of equation (5.6.1) that is valid in a neighborhood of the extremum of the function $\alpha(r)$. To evaluate the horizontal scale of the generated velocity field, the growth rate and the generation threshold, we make use of the asymptotic solution of the problem for the magnetic dynamo obtained by Sokolov *et al* (1983). Converted to the problem at hand, this corresponds to the solution of (5.6.1) for large values of the parameter $R_\alpha = \alpha_0 L_\alpha/\nu$, where L_α is the characteristic scale on which mean helicity varies and α_0 is its characteristic value. Consider the axisymmetric distribution $\alpha = \alpha(\rho)$ (ρ is the cylinder radius) with an extremum $\alpha = \alpha_0$ on the $\rho = 0$ axis. Then the steepest to grow is the velocity field which has the form (Moiseev *et al* 1983b)

$$\langle \boldsymbol{\Omega} \rangle = \frac{\Omega_0}{L_\alpha}\left\{0,\ \sin\frac{R_\alpha \rho}{2L_\alpha},\ \cos\frac{R_\alpha \rho}{2L_\alpha}\right\} \exp\left(\gamma_0 t - \frac{R_\alpha \rho^2}{4L_\alpha^2}\right) \tag{5.6.2}$$

close to the $\rho = 0$ axis (i.e. at $\rho \ll L_\alpha$). Here the increment γ_0 is

$$\gamma_0 \approx \frac{\alpha_0^2}{4\nu}\left(1 - \frac{4}{R_\alpha}\right); \tag{5.6.3}$$

we use the coordinate system $\{\rho, \varphi, z\}$. The velocity field (5.6.2) is helical: $(\langle \boldsymbol{v}\rangle \boldsymbol{\Omega}) \approx (R_\alpha/2L_\alpha)\langle v\rangle^2$. As follows from (5.6.2), the characteristic scale of vorticity decrease is $L = 2L_\alpha/\sqrt{R_\alpha}$. Using the data of Moiseev *et al* (1983b) we can write $\alpha \simeq 2\Omega_0 l \sin\Phi$. For $l \simeq 10$ km we find α $(\mathrm{m\,s^{-1}}) \approx \sin\Phi$, The characteristic scale of the area of TC inception is of the order of $L_\alpha \simeq 10^3$ km (Moiseev *et al* 1983b). For the turbulent viscosity coefficient we choose the conventional estimate $\nu \simeq 10^3\,\mathrm{m^2\,c^{-1}}$. This gives us

$$R_\alpha \approx 1.5\times 10^3 \sin\Phi \qquad L \approx \frac{50\,\mathrm{km}}{\sqrt{\sin\Phi}} \qquad T = \gamma^{-1} = \frac{0.5\,\mathrm{hours}}{\sin^2\Phi}.$$

For the latitude $\Phi = 10^\circ$ we find $L \approx 120$ km and $T \approx 1$ day, which is quite satisfactory.

It should be noted at this point that the amplification of vortical flows is only possible at sufficiently high values of the parameter R_α, that is, at $R_\alpha > R_{\alpha c}$ (see, for example, (5.6.3)). This is a confirmation of the familiar fact that tropical cyclones do not form at the equator. The upper bound on the latitude of typhoon formation appears to be caused by the temperature distribution in the surface layer of the ocean.

For a tropical cyclone to exist it is necessary that its central part contain upward flows, both in the northern and southern hemispheres, and since helicity α reverses its sign at the equator, the velocity fields must have opposing horizontal circulations. This is the explanation for the fact that typhoons are cyclonic in the northern hemisphere and anticyclonic in the southern hemisphere. Note finally that the component $\langle\Omega_\varphi\rangle$ in (5.6.2) reaches a maximum at a finite distance $\rho = 10$ km from the axis (for $\Phi = 10^\circ$). This peculiarity may explain the formation of the 'eye of the typhoon' during the nonlinear stage—the region at the center where tangential motion is insignificant. The estimate thus obtained agrees with the reported 'eye' radii within an order of magnitude (10 to 50 km). The mechanism proposed is thus able to ensure rapid exponential amplification (with a characteristic time of several days) of vortical perturbations on a scale of several hundred kilometers. The velocity field generated coincides both quantitatively and qualitatively with that typical of tropical cyclones.

Note that models in which large-scale perturbations play the main part but small-scale ones are ignored are quite popular (see, for example, Burpee and Reed 1982). The concept proposed by Moiseev *et al* (1983b) takes into account both classes of perturbation ($L_\alpha \gg L \gg 1$). To generate a vortex of size L, a large-scale disturbance on the scale L_α is required. However, small-scale perturbations play a critically important role, as we kept emphasizing earlier. It must be pointed out that we are not speaking of merely creating a feedback between toroidal and poloidal flows in tropical cyclones via small-scale perturbations. The fact is that this simultaneously reduces the role of viscous energy removal: the helical component of small-scale flow does not dissipate via viscosity to ever smaller scales; the opposite holds, i.e. a reverse cascade towards larger scales is generated. As a result this mechanism of TC generation offers advantages over other mechanisms.

Completing the analysis of this extremely simple model, we need to recall that helicity fluctuations act as negative viscosity (see section 5.2).

Now we will briefly discuss the possible results of explicitly taking convection into account, from the point of view of generation of TC-type large-scale helical structures. This is important for two reasons. First, as we have mentioned earlier, the regions of TC generation occur within regions of enhanced convection. Second, a small-scale turbulent ensemble is rather too weak an energy reservoir for such a powerful stucture as a typhoon; it needs efficient energy inflow, for which it can use the energy of thermal or, even better, humid convection.

An important role in studying the convective process is played by the nature of helical turbulence. We know that turbulence in tropical atmosphere produces very intense turbulent diffusion; it is difficult to say, however, whether this turbulence is definitely convective. Two versions were analyzed: (a) external large-scale helical turbulence with a large-scale convective process superposed on it (see, for example, Moiseev *et al* 1988a) and (b) convective helical turbulence that evolves on a locally unstable profile, the mean profile being neutral (see, for example, Lupyan *et al* 1992). According to the convective adaptation theory, it is precisely the convective turbulence that implements the tendency of the mean profile to neutrality (Khain and Sutyrin 1983).

Let us look at the first of these scenarios. We have analyzed it in section 5.2. The reader will recall that as helicity increases, the large-scale convection undergoes sudden restructuring: the horizontal size (in the uniform case) tends to infinity or (in the non-uniform case) a single helical cell is formed, of typhoon width $L \propto \sqrt{R_0 h}$ (h is the height of the atmosphere and R_0 is the radius of the heating area); $L \gtrsim 100$ km and the time of TC evolution is about 1 to 3 days. Taking humidity into account increases the increment for tropical parameters by a factor of 2 to 3.

In the second scenario (see, for example, Lupyan *et al* 1992) the final formulas are simpler because convection matches turbulence. The characteristic time of evolution of a TC at its earliest stage (tropical depression) is about ten days, and the size is about 80 km. If the process of helicity increase is not suppressed, the increment of the first mode abruptly starts to climb. The second mode is generated at the same stage. The two-mode structure fits real tropical storms quite well. Its evolution time is about one day.

Different convective regimes are thus in good agreement with the generation of large-scale helical vortices, their evolution times, sizes and topological characteristics. As for the energy aspects, their advantages are obvious, since the resources of thermal and latent energy in the tropics are vast. However, the two scenarios considered above greatly differ in the thresholds of large-scale structure generation. Let us summarize the differences. If only one small-scale helical turbulence exists, the threshold is determined by a single parameter $R_\alpha > R_{\alpha c} \approx 4$.

In the case of convection in the presence of external turbulence, the threshold is a function of two dimensionless parameters—the Reynolds number and the parameter proportional to the helicity of small-scale turbulence. As helicity increases, the threshold diminishes; it reaches its minimum as $k_\perp \to 0$ at a value of helicity above a certain critical value. The threshold also depends strongly on the type of boundary conditions. If free boundaries are replaced with heat-insulated ones, we obtain zero-threshold instability (Lyubimov *et al* 1991).

The type of threshold for tropical depression in the convection adaptation model is similar to the threshold of the previous scenario. However, the two-mode generation regime, whose type corresponds to a tropical storm, is rather bizarre, as far as the dependence on various parameters is concerned. For example, as the Taylor number $\mathrm{Ta} = 4(\Omega_1^2 - h^4)/\nu^2$ increases (h is the layer thickness and Ω_1 is the angular velocity of the Earth's rotation) and the friction with the driving layer decreases, the mode corresponding to the tropical depression decays faster as helicity increases than the system becomes ready for the formation of a rapidly evolving large-scale vortex.

It appears that all these scenarios can occur in real conditions, depending on which of the factors is favored. One of the tasks of early diagnostics of typhoon inception lies precisely in identifying the scenario of its evolution.

5.6.3 On typhoon precursors

The results outlined in the preceding sections make it possible to clarify the principally important aspects of typhoon forecast.

In what follows we briefly describe the qualitative side and also a number of experimental results related to using certain physical precursors of tropical cyclones. We need to lay special emphasis here on the particular role played by the satellite-based remote analysis in the study of atmospheric turbulence and large atmospheric vortices; this monitoring system provides continuous large-scale survey.

A. Fractal precursor

A method of early diagnostics of tropical perturbations was developed on the basis of infrared images of cloud layers produced by processing the data of meteorological satellites NOAA, GNS and Meteosat (Baryshnikova *et al* 1989). The method is based on analyzing the fractal dimension of temperature isolines in cloud clusters. The method can be used not only to predict the features of evolution of tropical perturbations but also for evaluating turbulence parameters. Several remarks are in order here on the idea of turning to fractals. Roughly speaking, the fractal dimension characterizes the extent of brokenness of a line or, in the same vein, of planar and three-dimensional objects. The dimension of the ordinary line is 1. The dimension of fractal curves lies between 1 and 2 and in this sense is an intermediate object between a line and a plane. The fractal curve is not a mere mathematical abstraction. Examples of fractal curves are the trajectory of a Brownian particle and the coastline curves. A temperature isoline in the atmosphere is also a fractal. Baryshnikova *et al* (1989) showed that the fractal dimension D of temperature isolines on IR-images of cloud cover depends on the properties of wind flows on which this cloud cover was formed. For a turbulent flow we have $D \gtrsim 1.3$ and for regular flows (ordered convection) $D \approx 1$. Fractal analysis showed that an abnormally high number of zones with ordered motion are observed in pre-typhoon situations. It is also important that temperature gradients at right angles to isolines are appreciably higher in ordered zones than otherwise. These results are physically quite clear. Since part of the turbulence energy is channeled to larger scales, transport coefficients grow smaller and, as a result, the fractal dimension decreases while the gradients across temperature isolines remain steep. The behavior of helicity fluctuations agrees with these results. As shown in section 5.2, they act as negative viscosity. Therefore we come to the opinion that the fractal precursor offers a powerful means of diagnostics of large-scale coherent processes in geophysics (not only in TC field).

B. Spectral precursors

Spectral peculiarities were studied using two types of instrument: RF-radiometers and Doppler radars.

First of all, we look at radiometric data, including temperature field fluctuations. The amplification of the reverse cascade at the phase when a tropical depression deepens is accompanied with changes in the spatial spectra of radio-brightness temperatures of the atmosphere (calculated from the data of a multichannel radiometric unit, Veselov *et al* 1989). An increase in slope of the spectra is in agreement with the theory; it predicts that energy is transferred from small-scale turbulence to large-scale structures.

Note now that a radiometric analysis of oceanic surface allows us to identify early anomalies of typhoon inception, as well as bends in the trajectories of mature typhoons (Guskov *et al* 1991).

Let us now turn to Doppler radar measurements. One important result was the identification of a knee on the spectrum of turbulent kinetic energy on the scale of tens of kilometers (Altaisky *et al* 1992).

Two features must be singled out. The first is the presence of turbulent pulsations on a scale greater than the region where the convective energy source is located; this argument favors the reverse cascade. Secondly, after the knee the spectrum is less steep than the Kolmogorov spectrum, which in all likelihood points to the helicity flux, not the energy flux, as the dominant one in this region.

Even though helical precursors were detected using non-satellite equipment (aircraft- and ship-borne), transplanting the methods to space satellite radiometric and radar instrument platforms should not involve problems of a principal nature. Some parameters will no doubt have to be changed.

C. Wave and admixture precursors

Different scenarios of wave perturbations are possible in a non-equilibrium medium (Moiseev *et al* 1984, Chimonas 1972). It is not surprising, therefore, that intense infrasound emission was detected prior to the birth of a tropical cyclone (Netreba 1991). The frequency range of the infrasonic emission is fairly broad: from 0.003 Hz to 10 Hz.

Infrasonic emission accompanied the transformation of the initial cloud cluster into a tropical depression. The evolution of this emission manifested its modulation with a period of about two hours. The further evolution of a tropical depression to the typhoon stage occurred after 2 to 4 surges of infrasonic noise, each lasting an hour and separated by intervals of about 40 hours. Even though this precursor is monitored by air- and ground-based instruments, a synchronous precursor is also generated in space, that is, surges of infrared thermal radiation.

Among the admixture precursors, we need first of all to single out changes in the total ozone concentration two to four days prior to the inception of a tropical cyclone and also in the vertical ozone distribution (Nerushev 1991). This precursor is monitored by both satellite-borne and non-satellite instruments.

An efficient multiparametric diagnostic of early typhoon stages has thus been achieved. It is very important that this diagnostic now has a solid scientific foundation.

5.6.4 On the mechanism of modulation of the number of strong earthquakes and of the Earth's rotation frequency

This section is based on the paper by Gokhberg *et al* (1995). We know that the spectrum of rotation frequency variations of the Earth shows characteristic periods from several hours to tens of years and even longer (Bucha 1983, Melchior 1986). It is typically assumed that short-period variations (shorter than or equal to 10 years) are caused by solar–lunar interactions, while the sources of variations with periods longer than ten years lie within the Earth.

Gokhberg *et al* (1995) have discovered that the 20-year cycle of repeatability of strong earthquakes of magnitudes $M \geqslant 6$ is similar to the variation of rotation frequency of the Earth with a period of $T \approx 20$ years.

A continuous series of the annual numbers of earthquakes of $M \geqslant 6.0$ over the interval 1918–1981 from the US Geological Survey earthquake catalog and the continuous series of annual differences between the ephemeris and solar days during the same interval were used as initial data (Gokhberg *et al* 1995). The same smoothing algorithm with a sliding interval of 5 and 11 years was applied to the two runs of observations. A linear correction of 0.05 ms/year for the secular deceleration of the Earth's rotation rate was applied to the smoothed run of day length deviations. A linear drift of -1 earthquake/year was excluded from the smoothed earthquake run. The resulting curves proved to be similar, with a period of $T \approx$ 22–25 years, and were very much in phase: a rise in the number of strong earthquakes coincides with the onset of the acceleration of Earth's rotation.

We will show in what follows that the processes unfolding in the Earth's liquid core and connected with the helical turbulence of compositional flows can produce additional local perturbations of pressure acting on the lower mantle; the characteristic spatial scale of the perturbations is several hundred kilometers. Those perturbations that reach up to the Earth's crust may, on the one hand, generate seismicity modulation in seismically active zones, and on the other hand, impose modulation on the frequency of Earth's rotation.

Compositional and thermal convection causes velocity fluctuations in the liquid core of the Earth (Braginskii 1963, Loper and Roberts 1983). It is currently accepted that the main source of motions in the liquid core is due to compositional convection (Loper and Roberts 1983). This convection results from iron crystallization on the surface of the inner core when the planet cools down, whereby the liquid ingredients enriched in the lighter component (sulfur released by crystallization) move up and float. The compositional convection in the core has to be turbulent because of the high Reynolds number (Loper and Roberts 1983). Owing to Earth's rotation, the liquid flows moving upwards from the hard inner core towards the mantle are twisted by the Coriolis force and gain helicity. The characteristic transversal and longitudinal sizes of these wall-shaped flows are 1 km and 1000 km, respectively, and the compositional convection is about $0.1\,\mathrm{cm\,s^{-1}}$ (Nikitina and Ruzmaikin 1991).

We have mentioned several times in this chapter that the small-scale turbulence possessing helicity may lead to generation of large-scale perturbations. This effect was used in section 5.6.2 to explain the inception of tropical cyclones in the atmosphere of the Earth. The equation that describes the generation of large-scale perturbations of velocity $\langle \boldsymbol{V} \rangle$ in the 'minimum-model' approximation has the form (5.6.1). This model is amply sufficient for a qualitative analysis at the current stage of investigation of the effect of modulation of the Earth's rotation frequency. We only need to introduce certain generalizations. First, angle brackets denote averaging over rotation period (of the Earth, in this case) or over the ensemble of realizations of a random field $\boldsymbol{V}^T$. Secondly, we take into account that the lifetime of a turbulent cell in real conditions may be less than the correlation time and assume that τ is either the lifetime of a turbulent cell or the correlation time. Do not forget that if $\alpha = \text{const}$, the growth rate γ of exponentially growing perturbations can be written as

$$\gamma = \alpha k - \nu k^2 \tag{5.6.4}$$

where k is the wave number. Equation (5.6.4) implies that there exists a characteristic perturbation scale

$$L = \frac{2\pi}{|k^*|}$$

at which the growth rate reaches its maximum value $\gamma^* = \alpha^2/(4\nu)$, where $k^* = \alpha/(2\nu)$.

According to Nikitina and Ruzmaikin (1991), for the Earth $\alpha = 10^{-4}\,\mathrm{m\,s^{-1}}$. The least certain quantity is the viscosity coefficient. If we assume a value of $\nu = 1.6\,\mathrm{m^2\,s^{-1}}$ that corresponds to averaged data (Lumb and Aldridge 1991), then we obtain $T \approx 20$ years and $L \approx 200$ km for $T = \gamma^{*-1}$ and L. This size L is of the order of the characteristic size of the area that is usually hit by strong earthquakes. Of course, a different value of ν produces different results for T and L. However, if the model suggested here is correct, it indirectly yields the characteristic value of the effective viscosity.

As pointed out above, large-scale velocity perturbations possess helicity (Moiseev *et al* 1983a, 1983b) and therefore, have a velocity component directed from the inner core to the mantle. Those perturbed flows of the liquid that reach the surface of the mantle interact with it, owing to the small-scale roughness of the outer core–mantle boundary (Melchior 1986, Toomre 1966). The transfer of the local torque from vortical liquid flows to the mantle via the linkage produces local twisting inside the mantle. The twisted mantle areas are squeezed upwards as a result of longitudinal momentum transfer from the rising vortices in the liquid core; this and some other favorable conditions[4] may drive the local upward transfer of mantle components, thereby increasing the pressure at the points of interaction between the mantle and the Earth's crust. The upward displacement of mantle material increases its moment of inertia and hence, slows down the rotation. The consequent sagging of the material that traveled upwards will reduce the moment of inertia and increase the rotation rate. Note that according to the results of processing the initial data and the model outlined above, the increase in the number of earthquakes correlates with the sagging of matter in the mantle.

Let us roughly evaluate the changes caused by the geometric characteristics of the displaced mass in the mantle rotation frequency. The angular momentum conservation condition,

$$I_m \Omega_m = \mathrm{const},$$

[4] Large-scale inhomogeneities at the outer core–mantle boundary can help the process of interaction between the rising vortices in the liquid core and the mantle constituents (Bemmelen 1966). These inhomogeneities lead to hyperdifferentiation of matter in the lower mantle and then to peridotite melting chambers (Belousov 1968) confined to zones of deep oceanic trenches of island arcs, mid-ocean ridges and other seismically active regions.

where I_m and Ω_m are the moment of inertia and mantle rotation frequency, implies

$$\frac{\Delta\Omega_m}{\Omega_m} = -\frac{\Delta I_m}{I_m}.$$

Here $\Delta\Omega_m$ is the change in the rotation frequency caused by the change in the moment of inertia by ΔI_m. The mantle's moment of inertia is

$$I_m = \int \rho_m r_s^2 \, dr$$

where ρ_m is the density of matter in the mantle and r_s is the distance of an element dr from the rotation axis. Assume that the lower base of the mantle column of cross-sectional area S (at the boundary of the external core) at a latitude φ has moved upwards along the radius by a distance h and the upper base (at the Earth's surface) remained fixed. Taking into account a compaction of matter as a result of this flow and also that the internal radius of the mantle is roughly $0.6\,R_E$, where R_E is Earth's radius, we find for P similar columns

$$\frac{\Delta I_m}{I_m} \approx \frac{5.6}{\pi} P \frac{\mathrm{Sh}}{\mathrm{R_E^3}} \cos^2\varphi. \tag{5.6.5}$$

We can assume for the sake of a crude evaluation that $P \approx N_{\mathrm{mod}}$, where N_{mod} is the difference between the maximal and minimal annual number of strong earthquakes within a 20-year cycle, that is, the maximum annual number of those that are related to mass transfer in the mantle treated by the hypothesis discussed here.

We can evaluate the right-hand side of (5.6.5) assuming $S \approx (\pi/4)L^2$, $L = 200\,\mathrm{km}$ (see above), $h = 1\,\mathrm{m}$, $\cos\varphi = 1/\sqrt{2}$, $N_{\mathrm{mod}} = 50$. This gives (for $R_E = 6375\,\mathrm{km}$) $\Delta I_m/I_m \approx 5.4 \times 10^{-9}$, which coincides with the quantity $|\Delta\Omega_m|/\Omega_m \approx 5 \times 10^{-9}$ returned by processing the initial data. The mass motions analyzed above can thus cause both strong earthquakes and appreciable changes in the rotation frequency of the Earth.

The periodicity of these processes in time can be explained in the following manner.

(1) In general, large-scale helical vortices in the liquid core must be born in large volumes within the core, that is, they must form a periodic spatial lattice of vortices, resembling the generation of periodic multilayer vortex structures in the atmosphere (Gryanik 1988). Such a process must be described by nonlinear equations for large-scale helical structures. Over a time of the order of inverse growth rate, that is, about 20 years in the case we now consider, helical vortices reach an amplitude sufficiently high to cause mass transfer on those parts of the mantle in which the conditions are most suitable. As mentioned above, such conditions are found in seismically active zones. In further evolution, vortices

may decay as a result of instability with respect to shear flow generation (Drake *et al* 1992). The time of evolution of this instability may be several times shorter than the time of formation of a helical instability if the vortical velocity is sufficiently high. This pattern may be periodically repeated.

(2) A simpler explanation is based on the dependence of helicity on the angular velocity of the Earth's rotation. Indeed, we saw above that the onset of a large-scale helical instability reduces the angular velocity, that is, reduces α. In its turn, this reduces the growth rate of helical instability and ultimately returns the system to its initial state, in response to the sagging of the uplifted mass of the mantle. The Earth's rotation rate increases again, and the growth rate and α climb up. The process will thus be periodic, as a result of the nonlinear mode of helical instability.

References

Altaisky M V, Klepikov I N and Moiseev S S 1992 *Annales Geophysicae* Suppl. I to vol 10, p 138

Artsimovich L A and Sagdeev R Z 1979 *Fizika Plazmy dlia Fizikov* (Moscow: Atomizdat) pp 1–307 (in Russian)

Baryshnikova Yu S, Zaslavsky G M, Lupyan E A, Moiseev S S and Sharkov E A 1989 *Issledovaniya Zemli iz Kosmosa* **1** 17–25

Batchelor G K 1953 *The Theory of Homogeneous Turbulence* (Cambridge: Cambridge University Press)

Belian A V, Chkhetiani O G and Moiseev S S 1996 *Preprint No 1957* (Moscow: Space Research Institute) pp 1–16

Belian A V, Moiseev S S and Petrosyan A S 1990 *J. Phys. Condens. Matter* **2** 469–75

Belousov V V 1968 *Zemnaya Kora i Verkhniya Mantiya Okeanov* (Moscow: Nauka) pp 1–255 (in Russian)

Bemmelen R W van 1966 *Tectonophys.* **3** 83–127

Berezin Yu A and Zhukov V P 1989 *Izv. Akad. Nauk SSSR MZhG* **4** 3–9

Braginskii S I 1963 *Dokl. Akad. Nauk SSSR* **149** 1311–4

Brissaud A, Frisch U, Leorat J, Lesieur M and Mazure A 1973 *Phys. Fluids* **16** 1366–7

Bucha V (ed) 1983 *Magnetic Field and Processes in the Earth's Interior* (Prague: Academia) pp 1–14

Burpee R W and Reed R J 1982 *GARP Publ. Ser.* **25**

Chimonas G 1972 *Bound.-Layer Met.* **2** 444–52

Drake J F, Finn J M, Guzdar P, Shapiro V, Shevchenko V, Waebroek F, Hassam A B, Liu C S and Sagdeev R 1992 *Phys. Fluids* **B4** 488–91

Druzhinin O A and Khomenko G A 1989 *Preprint No 1588* (Moscow: Space Research Institute) pp 1–21

Fortov V E, Gnedin Yu N, Ivanov M F, Ivlev A V and Klumov B A 1996 *Usp. Fiz. Nauk* **166** 391–422

Frisch U, She Z S and Sulem P L 1987 *Physica* **28D** 382–92

Gershuni G Z and Zhukhovitskii E M 1972 *Konvektivnaya Ustoichivost Neszhimayemoi Zhidkosti* (Moscow: Nauka) pp 1–192 (in Russian)

Gokhberg M B, Barsukov O M, Moiseev S S and Nekrasov A K 1995 *Dokl. Akad. Nauk* **341** 313–5

Gryanik V M 1988 *Izv. Akad. Nauk SSSR FAO* **24** 20–9

Guskov G Ya, Moiseev S S and Cherny I V 1991 *Preprint No 1762* (Moscow: Space Research Institute) pp 1–34 (in Russian)

Gvaramadze V V and Chkhetiani O G 1988 *Preprint No 1440* (Moscow: Space Research Institute) pp 1–37 (in Russian)

Gvaramadze V V, Khomenko G A and Tur A V 1987 *Preprint No 1210* (Moscow: Space Research Institute) pp 1–35 (in Russian)

Gvaramadze V V, Khomenko G A and Tur A V 1989 *Geophys. Astrophys. Fluid Dyn.* **46** 53–69

Hunt J C R and Carruthes D J 1990 *J. Fluid Mech.* **212** 497–532

Khain A P and Sutyrin G G 1983 *Tropicheskie Tsiklony i Ikh Vzaimodeistvie s Okeanom* (Leningrad: Gidrometizdat) pp 1–272 (in Russian)

Klyatskin V I 1975 *Statisticheskoe Opisanie Dinamicheskikh Sistem s Fluctuiruyushchimi Parametrami* (Moscow: Nauka) pp 1–239 (in Russian)

Klyatskin V I 1980 *Stokhasticheskie Uravnenia i Volny v Sluchaino Neodnorodnykh Sredakh* (Moscow: Nauka) pp 1–336 (in Russian)

Kraichnan R H 1973 *J. Fluid Mech.* **59** 745–52

Kraichnan R H 1976 *J. Fluid Mech.* **75** 657–75

Krause F and Radler K-H 1980 *Meanfield Magnetohydrodynamics and Dynamo Theory* (Berlin: Academic)

Krause F and Rudiger C 1974 *Astron. Nachr.* **295** 93–9

Landau L D and Lifshits E M 1976 *Statisticheskaya Fizika* Part 1 (Moscow: Nauka) pp 1–479 (in Russian)

Leslie D 1973 *Developments in the Theory of Turbulence* (Oxford: Clarendon)

Loper D E and Roberts P H 1983 *Stellar and Planetary Magnetism* (New York: Gordon and Breach) pp 297–327

Lumb L I and Aldridge K D 1991 *J. Geomag. Geoelectr.* **43** 9

Lupyan E A, Mazurov A A, Rutkevich P B and Tur A V 1992 *Zh. Exp. Teor. Fiz.* **102** 1540–52

Lupyan E A, Mazurov A A, Rutkevich P B and Tur A V 1993 *Dokl. Akad. Nauk SSSR* **329** 720

Lyubimov D V, Moiseev S S, Pungin V G and Cherny I V 1991 *EGS General Assembly Report OA10–16 (Wiesbaden, 1991)*

Markus P S and Press W H 1977 *J. Fluid Mech.* **77** 525–34

Melchior P 1986 *The Physics of the Earth's Core: an Introduction* (Oxford: Pergamon) pp 1–256

Merzliakov E G and Moiseev S S 1987 *Preprint No 1155* (Moscow: Space Research Institute) pp 1–18 (in Russian)

Moffat H K 1978 *Magnetic Field Generation in Electrically Conducting Fluids* (Cambridge: Cambridge University Press)

Moffat H K 1981 *J. Fluid Mech.* **106** 27–47

Moiseev S S, Sagdeev R Z, Tur A V, Khomenko G A and Janovsky V V 1983a *Sov. Phys.–JETP.* **58** 1144

Moiseev S S, Sagdeev R Z, Tur A V, Khomenko G A and Shukurov A M 1983b *Sov. Phys.–Dokl.* **28** 925–8

Moiseev S S, Suyazov N V and Etkin V S 1984 *Preprint No 905* (Moscow: Space Research Institute) pp 1–18 (in Russian)

Moiseev S S, Rutkevich P B, Tur A V and Yanovskii V V 1987 *Proc. Int. Conf. on Plasma Physics (Kiev, 1987)* vol 2, ed A G Sitenko (Kiev: Naukova Dumka) pp 75–9

Moiseev S S, Rutkevich P B, Tur A V and Yanovskii V V 1988a *Sov. Phys.–JETP* **67** 294–9

Moiseev S S, Rutkevich P B, Tur A V and Yanovskii V V 1988b *Plasma Theory and Nonlinear and Turbulent Processes in Physics* vol 2, ed N S Erokhin *et al* (Singapore: World Scientific) p 298

Moiseev S S, Tur A V and Khomenko G A 1988c *Preprint No 140* (Moscow: Space Research Institute) pp 1–34 (in Russian)

Moiseev S S, Oganian K R, Rutkevich P B, Tur A V, Khomenko G A and Yanovskii V V 1990 *Integriruemost i Kineticheskie Uravnenia dlia Solitonov* (Kiev: Naukova Dumka) pp 288–332 (in Russian)

Moiseev S S, Petrosyan A S, Sagdeev R Z, and Chkhetiani O G 1991 *Preprint No 1751* (Moscow: Space Research Institute) pp 1–28 (in Russian)

Molchanov S A, Ruzmaikin A A and Sokolov D D 1985 *Sov. Phys.–Usp.* **28** 307

Monin A S and Yaglom A M 1965 *Statisticheskaya Gidromekhanika. Mekhanika Turbulentnosti* (Moscow: Nauka) vol 1 pp 1–639, vol 2 pp 1–720 (in Russian)

Nemtsov B E and Eidman V Ya 1989 *Akustich. Zh.* **35** 882–6

Nerushev A F 1991 *Proc. V Int. Symp. on Tropical Meteorology (Obninsk, 1991)* (Obninsk: Institute of Experimental Meteorology) p 74 (in Russian)

Netreba S N 1991 *Proc. V Int. Symp. on Tropical Meteorology (Obninsk, 1991)* (Obninsk: Institute of Experimental Meteorology) p 103 (in Russian)

Nigmatulin R I 1987 *Dinamika Mnogofaznyh Sred* vol 1, 2 (Moscow: Nauka) (in Russian)

Nikiforov A F and Uvarov V B 1984 *Spetsialnye Funktsii Matematicheskoi Fiziki* (Moscow: Nauka) pp 1–344 (in Russian)

Nikitina L V and Ruzmaikin A A 1991 *Preprint No 21(968)* (Moscow: Inst. of Terr. Magnetism and Wave Prop.) pp 1–25

Ooyama K 1964 *Geophysica Intern.* **4** 187–98

Riel H 1976 *Climate and Weather in the Tropics* (London: Academic) pp 1–611

Rutkevich P B 1993 *Zh. Exp. Teor. Fiz.* **104** 4010–20

Rutkevich P B, Sagdeev R Z, Tur A V and Janovskii V V *Preprint No 1487* (Moscow: Space Research Institute) pp 1–17

Sagdeev R Z, Moiseev S S, Rutkevich P B, Tur A V and Yanovskii V V 1987 *Proc. Int. Symp. on Tropical Meteorology (Yalta, 1987)* (Leningrad: Gidrometizdat) pp 18–28

Sokolov D D, Shukurov A M and Ruzmaikin A A 1983 *Geophys. Astrophys. Fluid Dyn.* **25** 293–307

Steenbeck M, Krause F and Radler K-H 1966 *Z. Naturforsch.* **21a** 1285

Toomre A 1966 *The Earth-Moon System* (New York: Plenum) pp 33–45

Townsend A A 1976 *The Structure of Turbulent Shear Flow* (Cambridge: Cambridge University Press) pp 1–429

Tur A V, Khomenko G A, Gvaramadze V V and Chkhetiani O G 1987 *Proc. Int. Conf. on Plasma Physics (Kiev, 1987)* vol 2, ed A G Sitenko (Kiev: Naukova Dumka) pp 203–6

Vainshtein S N, Zeldovich Ya B and Ruzmaikin A A 1980 *Turbulentnoe Dynamo v Astrofizike* (Moscow: Nauka) pp 1–352 (in Russian)

Veselov V M *et al* 1989 *Preprint No 1604* (Moscow: Space Research Institute) pp 1–12 (in Russian)
Yokoi N and Yoshizawa A 1993 *Phys. Fluids* A **5** 464–77
Yoshizawa A 1992 *Phys. Rev.* A **46** 3292
Yoshizawa A and Yokoi N 1991 *J. Phys. Soc. Japan* **60** 2500

Chapter 6

The importance and prospects of further research into nonlinear instabilities in helicoidal media (concluding remarks)

6.1 Two-dimensional or helical turbulence?

We have demonstrated in the preceding chapter that hurricane-type large-scale structures in Earth's atmosphere can grow out of small-scale helical motions. However, the two-dimensional model of the atmosphere (i.e. one that neglects three-dimensional motions) must also contain a possibility of a similar process (see, for example, Starr 1968). It is therefore interesting to discuss whether a helical model is competitive in comparison with the two-dimensional turbulence model.

The process of 'spreading', of a growing scale, is fairly typical of a two-dimensional (or rather, quasi-two-dimensional) medium. In view of a relatively small thickness of the atmosphere, it may seem to be just such a medium, and it would in principle be possible to have a large-scale hurricane vortex formed within a two-dimensional model from small-scale motions. Actually a number of problems arise when this model is used. We know that hurricanes can develop very rapidly but that two-dimensional processes do not allow for that. However, changing to simple three-dimensional motions results in elimination of the tendency to scale growth.

The way out of this cul-de-sac is found in taking into account the helical nature of motions. Three-dimensional helical motions show a tendency to merge and produce larger structures. It is important that this process is coherent and 'collapse'-like in that it evolves exponentially fast. We also need to emphasize that such structures as hurricanes cannot be 'squeezed' into the two-dimensional frame, owing to their essentially three-dimensional topologically non-trivial nature.

Even this example shows that the helical model deserves special attention. It is also obvious that a model gains in advantages if it is a multi-purpose one. The helical model of turbulence is of certain interest in this respect: the spectral theory needs modification for the calculation of asymptotic spectral distribution of turbulent energy, and transfer phenomena involving vector quantities (momentum, electric current etc) need investigation. Helicity is a pseudoscalar and as a result the system characterized by this parameter has no symmetry center (a pseudoscalar reverses its sign under mirror reflection at the center). A media without symmetry center is know as chiral. Chiral media have the ability to generate structures. We can see a demonstration of this in the example of media with helical turbulence. The cause is the thermodynamic instability of chiral media. Thus if a distribution function describing the state of a medium depends on a chiral parameter, the system is undoubtedly thermodynamically unstable (compare with the equilibrium Gibbs function which is independent of pseudoscalars). The following are typical pseudoscalars that can characterize the chiral properties of the medium:

(i) Helicity (denoted in this case by S_V). This parameter is important in the MHD-dynamo and the vortical dynamo in non-conducting media described in the preceding chapter.

(ii) $S_H = \langle (\boldsymbol{h}' \cdot \operatorname{rot} \boldsymbol{h}') \rangle$ ($\boldsymbol{h}'$ is a small-scale random magnetic field). For example, the electron MHD dynamo is related to S_H.

(iii) $S_M = \langle (\boldsymbol{M}' \cdot \operatorname{rot} \boldsymbol{M}') \rangle$ ($\boldsymbol{M}'$ is the magnetic moment). S_M determines the generation of helicoidal superstructures in regular magnetics without a center of symmetry (see section 6.3).

(iv) $\boldsymbol{j} = \sigma_0 \boldsymbol{E} + \sigma_k \operatorname{rot} \boldsymbol{E}$ is the generalized Ohm's law for chiral media ($\boldsymbol{j}$ is the current, $\boldsymbol{E}$ is the electric field, σ_0 is the isotropic conductivity, σ_k is the chirality parameter). The electric dynamo is related to the pseudoscalar parameter σ_k. The self-generation of electric fields was treated by Belian *et al* (1997).

The helical model can thus be significantly extended on the basis of the mechanisms of self-organization of chiral media. It is clear that in addition to two-dimensional turbulence, one also needs to use the three-dimensional helical model as an alternative for analyzing complex phenomena in turbulent MHD flows. In this book we have paid considerable attention to this model not only because we tend to treat it as our favorite but also to eliminate the existing 'bias': very much attention is given to the two-dimensional model while only relatively few scientists investigate an important area of helical dynamics. In order to compare the advantages and disadvantages of these two ways of analyzing the data, let us summarize the basic physical phenomena that follow from the competing approaches.

We begin with the two-dimensional turbulence (see, for example, Henoch *et al* 1990). The properties of two-dimensional turbulence and transfer processes in flows with such turbulence in particular were theoretically studied by Batchelor (1969), Kraichnan (1967), Leith (1968), Lilly (1969) and others. A theoretical analysis has shown that the two-dimensional turbulence, or strongly anisotropic turbulence, is characterized by an inverse transfer of turbulence energy. This means that energy is transferred not from large vortices to smaller ones until it gets dissipated by viscous forces over small scales, as normally happens in ordinary three-dimensional turbulence. Instead, energy is transferred to larger vortices for which viscous dissipation is negligible. This process results, under constant turbulent pumping, in an inverse cascade of turbulent energy. Consequently, such flows allow long-lived two-dimensional large-scale structures to develop. Being two-dimensional, such structures do not contribute to momentum transfer but greatly enhance the transfer of passive scalar component and heat. This enhancement of transfer is obviously equally anisotropic.

Another spectacularly unique feature of anisotropically turbulent flows is the so-called negative turbulent viscosity. This means that the energy of turbulent motion is not extracted from the mean flow; far from it: it is transferred from large-scale turbulent fluctuations to the mean flux. Starr (1953) was perhaps the first to discover this phenomenon by analyzing the data on atmospheric flows. His conclusions were later supported by Monin (1956), Webster (1965), Ozmidov *et al* (1970) and others both for the atmosphere and the ocean.

The kinetic energy spectrum of two-dimensional turbulence has two asymptotics, $E(k) \sim k^{-5/3}$ and $\sim k^{-3}$, in view of the fact that in the non-dissipative case, energy and entropy are conserved (Kraichnan 1967).

For helical turbulence, which is in principal a three-dimensional phenomenon (even though the anisotropy of scales may be very large), we have to emphasize from the very beginning that the inverse transfer is its main property and, in contrast to the case of two-dimensional turbulence, the inverse transfer often evolves as exponentially fast generation of a large-scale structure (as instability). Negative viscosity, or rather a negative additional term in the viscosity, is also characteristic of helical turbulence, as will be shown in the next section. We have already pointed out in chapter 5 that the mean helicity of the small-scale turbulence results in a generation term in the equation for the mean velocity, which is of dissipative nature and depends on velocity derivatives; in fact this term resembles negative velocity. Fluctuations of mean helicity result, on the other hand, in a dissipative term in the equation for the mean viscosity, whose structure is completely identical to the viscous term but with a negative coefficient.

The models are thus practically interchangeable in their principal, robust features and a subtler analysis is required to differentiate between them. A criterion for this could, for example, be the type of asymptotics of the energy spectrum $E(k)$ which is known (see, e.g. Moiseev and Chkhetiani 1996) to scale as $k^{-7/3}$ and $k^{-5/3}$. In addition to $E(k)$, the important factors are the rate of evolution of the large-scale velocity field, the topology of the structure that develops, the functional behavior of the boundary between the asymptotics $E(k)$, etc. Answers to these questions will determine the final choice of the model. At the moment we will give an example of a successful application of the helical mechanism to explain a natural phenomenon: an analysis of processes induced by the encounter of the Shoemaker–Levy 9 comet and Jupiter; we follow the paper of Kamenets *et al* (1997).

6.2 Vortical processes in the Jovian atmosphere induced by the collision with the Shoemaker–Levy 9 comet

The collision of the Shoemaker–Levy comet with Jupiter in July 1994 offered a unique chance of monitoring the evolution of planetary-scale physical processes induced by a very powerful natural cataclysmic event. The Jovian gravitational field has broken the core of the comet, long before the collision, into more than 20 pieces, from several hundred meters to two kilometers in diameter, which aligned into a chain and entered the atmosphere at a velocity of about 65 km s^{-1}. This produced favorable conditions for observing the response of the atmosphere to powerful localized perturbations at various levels of energy release (from less than 10^{28} to 10^{31} erg).

Most of the effects recorded (a strong shock wave emerging in the upper layers of the atmosphere, intense ionospheric perturbations, formation of long-lived vortex structures etc) were predicted earlier (Klumov *et al* 1994, Kamenets *et al* 1994) and soon confirmed (Hammel *et al* 1995). This has led to a sufficiently elegant model of the interaction between the fragments of the comet and the atmosphere, beginning with the initial braking phase, explosion of the fragment in the dense layers of the atmosphere and ending with the evolution of the trace that was created in the upper layers, driven by the Coriolis force and zonal winds.

This picture could not explain, however, both the very large diameters and the observed structures of the initial perturbations in the Jovian atmosphere that were evident during the first hours after the collision of each fragment of the comet with Jupiter. According to observations by the Hubble Space Telescope (Hammel *et al* 1995), all traces generated by the largest fragments had distinct common features that were continuously observed both in the visual (336 and 555 nm) and infrared (889 nm) ranges.

In the visual range we see a dark homogeneous central spot surrounded by an excentric ring which forms the inner boundary of an annular perturbed region whose outer boundary is not as well defined. In the IR range, the central spot is lighter and stands out against the darker surrounding background. The spatial scale of characteristic features in the zone where the largest fragments had hit is reported by Hammel *et al* (1995) and is practically the same for all fragments. The central ring radius reaches 3000 km; its expansion velocity was 450 ± 20 m s^{-1} and remained constant as the process evolved. The distance between the outer boundary of the perturbation region and the center of the inner spot was 12,000 km. This structure of the perturbation zone was observed for several hours after the fall of each of the larger fragments. The fine structure would then become invisible and the collision area was marked by a dark spot that was gradually stretched along the corresponding latitude.

An analysis of the dynamics of ring-like structures detectable in the traces left by large comet fragments has shown (Ingersol and Kanamori 1995) that the expansion of the region occupied by the perturbation is not caused by the transfer of the comet's particles in the upper atmospheric layers and is essentially a wave process. A comparison of the brightness of structures in various spectral ranges warrants a hypothesis that they were formed as a result of modulation of the upper edge of the clouds. Ivanov *et al* (1996) and Fortov *et al* (1996) were able to show that the wave processes generated by powerful energy releases in the atmosphere, namely the acoustic, gravitational and shock waves, could not provide the mechanism of the observed formation of perturbation in the cloud layer. An analysis of the mechanisms described above has shown, among other things, that the energy consumption for the generation of perturbations of the type outlined above is very high and greatly exceeds the level of energy input even by the largest fragments. Therefore, in order to interpret the picture observed, one has to take into account an additional energy release channel.

This additional energy source may be the energy of convective motions stored in the Jovian atmosphere, provided we accept that the conjecture of the hurricane nature of the observed traces is correct. We assume, in accordance with the work cited above, that the shock wave and the rising fireball formed by the explosion of the fragment only act as a trigger mechanism for the generation of a vortex process in the atmosphere of Jupiter. Furthermore, we apply the approach developed by Moiseev *et al* (1983) and postulate that the main energy source for the generation of an intense vortex is connected with vertical thermal convection whose considerable intensity is typical of Jupiter (Chamberlain 1978). The vertical temperature distribution shows that the most intense convective transfer occurs in the part of the troposphere under the tropopause ($p = 0.1$ atm) which spans the layers from the deepest to the upper boundary of the cloud layer. The characteristic vertical scale of convective cells in this case is around 100–150 km (Keller and Yarovskaya 1976).

Let us separate the field of flow caused by the effect of the thermic on the surrounding atmosphere into quasi-regular large-scale structures with characteristic velocities $\langle \boldsymbol{V} \rangle$ (angle brackets indicate averaging over the ensemble of realizations of the field $\boldsymbol{V}$) and small-scale perturbations, which also include turbulent convective cells. We will also make use of the results of section 5.6.2 to describe the evolution of mean vorticity $\boldsymbol{w} = \mathrm{rot}\langle \boldsymbol{V} \rangle$, applying equation (5.6.1) for $\boldsymbol{w}$:

$$\frac{\partial \boldsymbol{w}}{\partial \boldsymbol{t}} = \mathrm{rot}(\alpha \boldsymbol{w}) + \nu \Delta \boldsymbol{w} \tag{6.2.1}$$

where ν is the coefficient of turbulent viscosity, $\alpha \simeq \frac{\tau}{3}\langle \boldsymbol{V}\,\mathrm{rot}\,\boldsymbol{V} \rangle \simeq 2\Omega l \sin \Phi$ is the coefficient that characterizes the helicity of convective turbulence ($\boldsymbol{V}$ is the velocity field of convective turbulence, Ω is the angular velocity of rotation of the planet, l is the characteristic scale of a convective cell, Φ is the geographic latitude and τ is the correlation time).

Analyzing the basic physical parameters that determine the formation and evolution of the trace, we single out three phases in its evolution. A high-intensity vortex forms because the rising cloud of hot gas sucks in and rotates the surrounding atmosphere at the initial phase of growth of a large-scale structure in the region of fall of a large cometary fragment; this vortex plays the role of the main perturbation factor. Further evolution of this central perturbation is dictated by its further enhancement owing to the inflow of additional energy from convective motions in the atmosphere. The final stage of formation of a trace in the atmosphere is determined by the effect of the latitude-dependent Coriolis force on the atmospheric wind (by this time the vortex size becomes comparable with the Rossby radius) and by horizontal atmospheric flows which spread the generated perturbation in the atmosphere.

To evaluate the characteristic scale of the perturbation formed at the first stage of the process, we use the asymptotic solution of equation (6.2.1) which can be written in cylindrical coordinates, making use of the expressions (5.6.2) and (5.6.3), as

$$w = \frac{w_0}{L_\alpha}\left\{0, \sin\frac{R_\alpha r}{2L_\alpha}, \cos\frac{R_\alpha r}{2L_\alpha}\right\}\exp\left(\gamma_0 t - \frac{R_\alpha r^2}{4L_\alpha^2}\right) \qquad (6.2.2)$$

where L_α is the characteristic scale of helicity variation, $R_\alpha = \alpha_0 L_\alpha/\nu$ is a dimensionless parameter, α_0 is the maximum value of α at $r = 0$, and γ_0 is the growth rate of the solution,

$$\gamma_0 \approx \frac{\alpha_0^2}{4\nu}\left(1 - \frac{4}{R_\alpha}\right). \qquad (6.2.3)$$

There are two latitudinal zones in the Jovian atmosphere, having qualitatively different convection (Chamberlain 1978): the equatorial ($0 < \Phi \leqslant 40°$) and the midlatitude zone ($40° < \Phi \leqslant 60°$). The width of the midlatitude zone that covers the region where the cometary fragments had fallen is about 20° of latitude, or 24,000 km. We are of the opinion that this scale is a natural choice for evaluating L_α. Keller and Yarovskaya (1976) have analyzed the shear and convective flows in the atmosphere of Jupiter and obtained the estimate $\nu \sim 10^{10}\,\mathrm{cm^2\,s^{-1}}$. The origin of the cylindrical frame of reference in which the maximum helicity coefficient is reached must be related to the initial large-scale vortex that is formed when the thermic passes through the cloud layer. Using then the results of numerical simulation of a thermic's rise (Klumov *et al* 1994) we assume $l \sim 100$ km. With these characteristic values of l, L_α and ν, we evaluate the coefficients in equations (6.2.2) and (6.2.3) and arrive at $\alpha \approx 25$ m s^{-1} and $R_\alpha \approx 60$.

As follows from (6.2.3), a vortex can grow only if $R_\alpha > R_\alpha^{cr} = 4$. In this particular case this condition is satisfied, and hence the vortical motion caused by the thermic should be enhanced.

The region of growth of vortical perturbations is bounded by the condition $r \leqslant [v(R_\alpha - R_\alpha^{cr})]^{1/2}$; by the time $t \approx 2$ hours after the fragments have hit the atmosphere, its radius reaches the size $r_0 \approx 2000$ km. This region was logically identified on photographs of traces on the surface of Jupiter with the central dark ring. According to Hammel *et al* (1995), its radius by the corresponding time moment was, for different fragments, from 1788 to 2260 km. For earlier moments of time, the ring radii calculated using the formula given above also agreed, within measurement error, with the data of Hammel *et al* (1995). We can thus assume that the accuracy of this formula is not lower than the accuracy of the corresponding formulas suggested by Hammel *et al* (1995).

The central ring has maximum brightness in spectral bands that correspond to the emission bands of methane. It can therefore be interpreted as the zone of gas outflow from deeper layers to the upper edge of the cloud layer. This interpretation is confirmed by the reddish–brown coloring of this area, as photographs in natural colors have revealed (Hammel *et al* 1995). We know (Marov and Kolesnichenko 1987) that the upper layers of Jovian clouds are formed of ammonia; they look light owing to glittering white ammonia crystals. Deeper cloud layers contain ammonia hydrosulphate whose crystals are reddish–brown. When these layers are lifted to the top, they form reddish–brown areas that stand out clearly against the light background of ammonia clouds.

The asymptotic solution (6.2.2) is not applicable for the evaluation of vortical perturbations at subsequent stages. However, this estimate can be obtained by analyzing the solution for a large-scale vortex that evolves against the background of turbulent cells with constant helicity coefficient (Moiseev *et al* 1983). In this case the characteristic perturbation radius is given by the relation $r_1 = 0.5L_\alpha/x_\alpha$ where x_α is the first root of the zero-order Bessel function. For the chosen value of L_α, we find $r_1 \approx 4800$ km. This value is in good agreement with the maximum radius of well defined ring-like structures, 4649 km, observed on the photographs of the traces produced by large fragments of the comet (Hammel *et al* 1995).

The observed radius of the whole area occupied by the perturbation in the first hours of formation of a trace, $r_2 \approx 12{,}000$ km, coincides with the estimate of the characteristic scale of the midlatitude convective zone obtained by analyzing the types of thermal convection in the atmosphere of Jupiter. We can assume, therefore, that the vortical perturbation induced by the fall and explosion of a large cometary fragment rapidly involves the entire region with only one type of convective flow. According to Moiseev's work (1990), the characteristic vertical scale H of a hurricane is found from $H = h/\pi$, and in the case in hand is 30 to 50 km. As follows from the estimates given above, the thickness of the atmospheric layer occupied by the perturbation becomes comparable with the thickness of the cloud layer already at the initial stage of the process, and the horizontal size exceeds the vertical size by two orders of magnitude. This process must result in intense vertical mixing, leveling of velocities and temperatures at different altitudes and in the destruction of quasi-stationary convective cells. As a result, energy influx to a large-scale vortex must peter out by the end of the first day after its formation, in contrast to terrestrial hurricanes that get energy from the underlying surface of the ocean for a very considerable length of time.

Note that the generation threshold for a large-scale vortex is determined by the inequality $\alpha H/\nu > \pi$ (Moiseev *et al* 1983). According to the estimates given above, $H = 30$ km, and hence the constraint on the helicity coefficient is $\alpha \geqslant 100$ m s^{-1}, which gives us the constraint on the size of a convective cell, $l \geqslant 400$ km. For the Jovian atmosphere this condition appears to be quite realistic, and is confirmed independently by the mentioned estimates of cell sizes derived from the evaluation of the turbulent viscosity coefficient (Keller and Yarovskaya 1976).

Let us discuss now how the model chosen here reflects the real features of the helical scenario. We need to stress that equation (6.2.1) describes its main feature, namely, the relation between the degrees of freedom, which arises precisely because of helicity. This equation also takes into account the inhomogeneity of the atmosphere. If the feedback is simulated correctly, equation (6.2.1) has, as pointed out above, unstable solutions that describe the large-scale structure generation; this is the most important feature for our analysis. As for the form of this equation itself, this form of (6.2.1) is obtained if the compressibility of the atmosphere or its multicomponent composition is included in the model (Moiseev 1990). A detailed analysis of thermal convection and small-scale helical turbulence generates a still more complex model. However, the main feature, that is, the topological nontriviality (linkage of streamlines) is retained. As a result, the growth increment and the characteristic size of the system generated are found to be close, within an order of magnitude, to the values obtained using a simpler model (6.2.1). As for the instability thresholds, the following general tendency is found: a stronger decrease of the threshold is observed in a more realistic situation (polluted atmosphere, inclusion of heat exchange between the turbulent zone and the surrounding area, inclusion of helicity fluctuations etc) than follows from an analysis of equation (6.2.1). Obviously, the range of applicability of the helicity mechanism could then be extended still further.

Papers published after the paper by Moiseev *et al* (1983) showed that in addition to its ability to generate large-scale structures, the helical turbulence possesses another important property, namely, it can reduce turbulent viscosity. This tendency can be easily derived from the model (6.2.1) either by adding fluctuations of the helicity parameter α (these fluctuations may be anomalously large at the stability threshold) or when taking into account the higher correlational moments in the non-Gaussian approximation (Belian *et al* 1994). Let us first consider the former, simpler variant (see section 5.2 and Moiseev 1990). Let helicity vary in time with a characteristic time scale τ_1, $\tau \ll \tau_1 \ll \tau_2$ where τ and τ_2 are the correlation time of the helical component of the turbulent velocity field and the characteristic time of generation of the large-scale structure. Applying the assumptions made above and assuming that the helicity coefficient

α is independent of space coordinates, we can express it as a sum of the time-independent mean value α_0 and a statistically stationary additional term $\alpha'(t)$ with zero mean value, which fluctuates at a negligibly small correlation time:

$$\alpha = \alpha_0 + \alpha'(t) \qquad \langle\alpha'(t)\alpha'(t')\rangle = 2D\delta(t - t').$$

To simplify the averaging procedure, we assume that fluctuations are Gaussian. Averaging then (6.2.1) over the ensemble of realizations $\alpha'(t)$ we obtain (Moiseev 1990, Klyatskin 1975)

$$\frac{\partial}{\partial t}\langle \boldsymbol{w}\rangle = \alpha_0 \operatorname{rot}\langle \boldsymbol{w}\rangle + (\nu - D)\Delta\langle \boldsymbol{w}\rangle. \tag{6.2.4}$$

The form of the coefficient with the Laplacian in the right-hand side of equation (6.2.4) shows that the effective viscosity coefficient diminishes, which reduces the instability threshold due to helicity fluctuations. A visually clear interpretation of the drop in viscosity can be obtained by qualitative arguments. Indeed, it is easily shown that helicity enhances the tendency of vortices to merge. However, viscous damping is weaker for larger vortices, that is, they live longer and, on average for the ensemble, the effective viscosity coefficient decreases. In addition to a decrease in viscosity, pressure also diminishes in helical turbulence. Indeed, using Euler's equations, we find

$$\frac{\partial \boldsymbol{V}}{\partial t} - [\boldsymbol{V} \operatorname{rot} \boldsymbol{V}] = -\nabla\left(\frac{P}{\rho} + \frac{V^2}{2}\right)$$

$$\frac{\partial}{\partial t} \operatorname{rot} \boldsymbol{V} = \operatorname{rot}[\boldsymbol{V} \operatorname{rot} \boldsymbol{V}].$$

Let us consider an ensemble of small-scale helical motions with $\langle \boldsymbol{V}\rangle = 0$, limiting the analysis to the case of Beltrami flow, when $\operatorname{rot} \boldsymbol{V} = \lambda \boldsymbol{V}$, where λ is a parameter. In this case Euler's equation yields the Bernoulli equation

$$\left\langle\frac{P}{\rho}\right\rangle + \left\langle\frac{V^2}{2}\right\rangle = \text{const}$$

with a common constant for the entire region of helical motion. In other words, as the energy of the helical motion increases, the pressure decreases. We can expect that this effect qualitatively survives for other types of helical motion as well, provided $\langle(\boldsymbol{V} \operatorname{rot} \boldsymbol{V})\rangle \neq 0$. A bounded region with helical motions thus shows a tendency to self-organization.

Let us now consider the evolution of a vortex at the final stage, after the helical mechanism has been saturated and the further evolution of the resulting large-scale vortex is dictated by its interaction with strongly inhomogeneous zonal winds in Jupiter's atmosphere. We take into account the horizontal baroclinicity of the vortex and also the turbulent viscosity and heat conduction. The equations of motion, continuity and heat balance for a shallow rapidly rotating two-dimensional atmosphere (Kamenets *et al* 1997) will serve as the starting point:

$$\frac{\mathrm{d}\boldsymbol{V}}{\mathrm{d}t} = \frac{1}{\rho}\nabla_\perp P + f[\boldsymbol{V}, \boldsymbol{e}_z] + \nu\Delta_\perp \boldsymbol{V} \qquad (6.2.5)$$

$$\frac{\partial\rho}{\partial t} + \operatorname{div}\rho\boldsymbol{V} = 0 \qquad (6.2.6)$$

$$\frac{\mathrm{d}P}{\mathrm{d}t} + \gamma P \operatorname{div}\boldsymbol{V} - \bar{k}\Delta_\perp T = 0 \qquad \bar{k} = \frac{kR}{c_v}. \qquad (6.2.7)$$

Here $\nabla_\perp = \boldsymbol{e}_x(\partial/\partial x) + \boldsymbol{e}_y(\partial/\partial y)$; $\Delta_\perp = \nabla_\perp^2$, $\boldsymbol{e}_x$, $\boldsymbol{e}_y$, $\boldsymbol{e}_z$ are Cartesian unit vectors of which $\boldsymbol{e}_x$ and $\boldsymbol{e}_y$ point to the east and the north, respectively, and $\boldsymbol{e}_z$ is along the local vertical line, $\boldsymbol{V} = (V_x, V_y)$ is velocity, P, ρ and T are pressure, density and temperature, respectively, $f = f_0 + \beta_y$ is the Coriolis parameter in the β-plane approximation, $\gamma = c_p/c_v$ where c_p and c_v are the specific heat at constant pressure and at constant volume, respectively, ν and k are the effective kinematic viscosity and heat conduction coefficients, R is the gas constant, and $\mathrm{d}/\mathrm{d}t = \partial/\partial t + (\boldsymbol{V}\nabla)$. It is assumed that viscosity and heat conduction are small and $P = R\rho T$. Introducing the potential temperature θ by

$$\theta = T\left(\frac{P_0}{P}\right)^{\frac{\gamma-1}{\gamma}} \qquad (6.2.8)$$

we can derive from (6.2.6) and (6.2.7) an expression (Pedlosky 1987)

$$\frac{\mathrm{d}\theta}{\mathrm{d}t} = \left(\frac{P_0}{P}\right)^{\frac{\gamma-1}{\gamma}} \kappa\Delta_\perp T \qquad \kappa = \frac{k}{\rho c_p}$$

where κ is the temperature conductivity.

Since we only consider slow motions, the expression for velocity $\boldsymbol{V}$ in equation (6.2.5) can be expanded in a series in a small parameter $(1/f)\mathrm{d}/\mathrm{d}t$. We assume

$$a = \{P, \rho, T, \theta\} \qquad a = a_0 + a' = a_0 + \bar{a}(y) + \tilde{a}(x, y, z) \qquad a_0 \gg a'$$

where $\bar{a}$ describes the inhomogeneity of the equilibrium atmosphere and $\tilde{a}$ is the wave component of a. Equations (6.2.6) and (6.2.7) then give, if we retain only the nonlinear terms of lowest order,

$$\frac{\partial P}{\partial t} - r_R^2\frac{\mathrm{d}_0}{\mathrm{d}t}\Delta_\perp P' - V_R^*(y)\frac{\partial\tilde{P}}{\partial x} - V_{R0}\frac{\tilde{T}}{T_0}\frac{\partial\tilde{P}}{\partial x} + \frac{c_s^2}{\rho_0 f_0}J(P', \Theta')$$
$$+ \nu r_R^2\Delta_\perp^2\tilde{P} - \frac{\bar{k}}{k}\frac{\mathrm{d}_0}{\mathrm{d}t}\Theta' = 0 \qquad (6.2.9)$$

$$\frac{\mathrm{d}_0}{\mathrm{d}t}\Theta = \kappa\Delta_\perp\tilde{T} \qquad (6.2.10)$$

where

$$\frac{\mathrm{d}_0}{\mathrm{d}t} = \frac{\partial}{\partial t} + \mathbf{V}\nabla \equiv \frac{\partial}{\partial t} + \frac{1}{\rho f} J(P, \ldots) \qquad J(a,b) = \frac{\partial a}{\partial x}\frac{\partial b}{\partial y} - \frac{\partial a}{\partial y}\frac{\partial b}{\partial x}$$

$$V_R(y)^* = V_R(y) + V_{R0}\frac{\bar{T}}{T_0} \qquad V_R(y) \approx V_{R0}\left(1 - a\frac{r_R}{r_f}y\right)$$

$$V_{R0} = \frac{c_s^2}{f_0^2}\beta_0 \qquad r_R^2 = \frac{c_s^2}{f_0^2} \qquad r_f = \frac{f_0}{\beta}$$

$$A = 2 + \tan^2 \Phi_0 \qquad c_s^2 = \gamma\frac{P_0}{\rho_0}$$

where r_R is the Rossby–Obukhov deformation radius, $V_R^*(y)$ describes the latitudinal inhomogeneity of the Rossby velocity, $V_R(y)$ corresponds to the local inhomogeneity of the Rossby velocity due to the dependence of $\beta(y)/f^2(y)$ on y, c_s is the local adiabatic speed of sound, $\mathbf{V}_g$ is the geostrophic velocity, and Φ_0 is the local latitude. The term $V_{R0}(\bar{T}/T_0)\partial\tilde{P}/\partial x$ corresponds to the scalar nonlinearity and the term $\frac{c_s^2}{\rho_0 f_0}J(P', \Theta')$ to the effect of baroclinicity.

The equilibrium pressure gradient $\bar{P}(y)$ in the geostrophic approximation is related to the zonal wind velocity $\bar{V}_x(y)$ by the formula

$$\bar{V}_x(y) = \frac{1}{\rho f}\frac{\partial \bar{P}(y)}{\partial y}. \tag{6.2.11}$$

Introducing dimensionless variables

$$P' \to \gamma P_0 p\frac{r_R}{r_f} \qquad \Theta' \to \theta\Theta_0\frac{r_R}{r_f} \qquad T' \to T_0 T\frac{r_R}{r_f}$$

$$(x, y) \to (x, y)r_R \qquad t \to t\frac{r_f}{c_s}$$

we find from equations (6.2.8)–(6.2.10) simplified equations for the horizontal inhomogeneous atmosphere (in dimensionless variables),

$$\frac{\partial p}{\partial t} - \frac{\mathrm{d}_0}{\mathrm{d}t}\Delta_\perp p - \left(1 - Ay\frac{r_R}{r_f}\right)\frac{\partial p}{\partial x} - (\bar{T} + \tilde{T})\frac{r_R}{r_f}\frac{\partial p}{\partial x} + J(p, \theta)$$
$$+ \bar{\nu}\Delta_\perp^2 p - \frac{\mathrm{d}_0\theta}{\mathrm{d}t} = 0 \tag{6.2.12}$$

$$\frac{\mathrm{d}_0\theta}{\mathrm{d}t} = \bar{\kappa}\Delta\tilde{T} \tag{6.2.13}$$

$$\tilde{T} = (\gamma - 1)\tilde{p} + \tilde{\theta} \tag{6.2.14}$$

where

$$\frac{\mathrm{d}_0}{\mathrm{d}t} = \frac{\partial}{\partial t} + J(p, \ldots) \qquad \bar{\nu} = \nu\frac{r_f}{r_R^2 c_s} \qquad \bar{\kappa} = \kappa\frac{r_f^2}{f_R^3 c_s}\frac{T_0}{P_0}.$$

Introducing the generalized vorticity

$$q = \Delta_{\perp} p - p + \frac{r_R}{r_f}\Psi + y - Ay^2 \frac{y^2}{2}\frac{r_R}{r_f} + \theta$$

where

$$\Psi(y) = \int_0^y \bar{T}(y')\, dy'$$

we rewrite equation (6.2.12) in the form

$$\frac{d_0}{dt} q = J(p, \theta) - \tilde{T}\frac{r_R}{r_f}\frac{\partial \tilde{p}}{\partial x} + \bar{\nu}\Delta_{\perp}^2 p. \tag{6.2.15}$$

The simplified equations (6.2.13)–(6.2.15) enable one to analyze the evolution of vortical structure in shallow baroclinic atmosphere in the intermediate geostrophic approximation. These equations can be treated as a generalization of the equations derived in Kamenets *et al* (1993, 1994) to which they reduce if we ignore the effects of viscosity and heat conduction and assume linear dependence of hydrodynamic parameters a on the coordinate y (when the zonal flow velocity is constant).

The set of equations (6.2.13)–(6.2.15) was numerically solved in the range $(x, y) \leqslant 20$ which lies in the mesosphere of Jupiter at a latitude of 45°. The size of the area was sufficiently large to eliminate the effect of boundaries. The initial perturbation was given by $a(r) = A\exp(-r^2/R^2)$ where R is the characteristic size and A is the perturbation amplitude. The values of the large-scale gradient of equilibrium pressure (and of its analog quantity, the gradient of potential temperature) were given in accordance with the observed latitudinal profile of zonal winds in the Jovian atmosphere (Kamenets *et al* 1996) in the framework of the quasi-geostrophic approximation (6.2.11). The values of the dimensionless viscosity and thermal conduction coefficients were taken to equal 0.002. The non-perturbed values of pressure, temperature and density were chosen to be $P_0 = 0.1$ bar, $T_0 = 110$ K and $\rho_0 = 10^{-5}$ g cm^{-3}.

Numerical simulation was carried out for two values of perturbation amplitude of pressure and temperature: $A_p = 0.01$, $A_T = 0.3$ and $A_p = 0.03$, $A_T = 0.3$. Assuming the altitude scale in the Jovian atmosphere to be 75 km, we arrive at energy estimates for the weak and strong perturbations to be 3×10^{28} and 10^{29} erg, respectively. In both cases the evolution of initial perturbations resulted in the formation of anticyclone structures which slowly shifted southward; for 20 to 30 terrestrial days after the explosion they struggled through an area unfavorable to anticyclone vortices (non-uniform zonal winds there countertwist the vortex in the direction opposite to that of its rotation). A vortex generated by a weak perturbation would decay in this area, leaving behind only a string of weak vortices, while a strong perturbation would only slightly change its amplitude, increasing its characteristic size and spawning a string of small vortices. By 55° of southern latitude, a strong vortex would stop its southward drift, having reached by then a size of 4×4 Rossby radii; its center would remain at the latitude where the zonal wind velocity reverses sign. This vortex would not decay during further motion, which can be understood in terms of compensation of dissipative losses by zonal winds whipping up the rotation of the vortex. The vortex energy in this area was about 10^{30} erg.

Numerical simulation thus showed that Jupiter's atmosphere is characterized by an energy threshold for initial perturbation, determined by the latitudinal velocity distribution of zonal wind and estimated for the case in hand as $\sim 10^{29}$ erg. A strong atmospheric perturbation with energy equal or greater than the threshold value transforms into a baroclinic anticyclone vortex whose size and energy evaluate close to the corresponding parameters of the Great Red Spot on Jupiter. The results of numerical simulation for strong perturbations are in good agreement with observations from the Hubble Space Telescope (Hammel *et al* 1995). The vortex evolution at the final stage is thus determined by the energy that it possesses in the field of saturation of the helical amplification mechanism.

In its turn, this energy depends, as does the size of the vortex amplified by the helical mechanism, on the parameters of the convection zone. For example, since the equatorial convective zone in the Jovian atmosphere has twice as large a latitudinal extension than the mid-latitudinal zone (for which the estimates above were obtained), we can expect that if the comet were to have hit closer to the equator ($0 < \Phi \leqslant 40°$), the diameter of the resulting vortex would have reached about 20,000 km by the end of the first terrestrial day, and 20,000 to 25,000 km by the end of the first month as a result of interaction with zonal winds. According to the work by Nezlin and Snezhkin (1993), such a vortex size corresponds to the criterion that guarantees the formation of a stable Rossby soliton, and is also close to the Great Red Spot diameter. This leads to a conjecture that if a comet with parameters close to those of the Shoemaker–Levy 9 comet fell in the equatorial convection zone of Jupiter's atmosphere, another long-lived vortical structure could have been generated, similar to the Great Red Spot which is indeed observed in the convective zone.

6.3 Some yet unsolved problems

To complete this chapter and the book as a whole, we would like to point out a number of important specific features of further progress in, first of all, the research effort aimed at nonlinear instabilities in chiral (mirror-noninvariant) media, and also in some related fields.

1. We begin with the 'related' aspect of the optimal evolution of a system. The instabilities treated in this book make it possible to formulate the common principle of evolution of physical systems with non-equilibrium sources. On the qualitative side, it can briefly be put as follows: a system tends at maximum speed to the maximum equilibrium of states that are possible under given conditions (see sections 2.1 and 3.1 and the paper by Moiseev and Pungin

1996). In this generalized form, this principle shows itself most vividly in the case of helicoidal magnetic structures, both in the rate of evolution (with maximum speed) and in the final state that sets in (and corresponds to the minimum of potential energy). It is also important that in this case the effects of the general properties of coherent generation of helicoidal structures are revealed in a relatively simple way.

Which of these common properties are the main ones? First, 'helical' terms of the type of $\boldsymbol{A}\,\mathrm{rot}\,\boldsymbol{A}$ (where $\boldsymbol{A}$ stands for one of the investigated or given fields) are invariably present in the principal functions, that is, the correlation function for magnetic and velocity fields, and the free energy for magnetics. One of the consequences of this property is that the generation of a coherent structure is possible both in the laminar and the turbulent cases. The important feature, therefore, is not whether averaging has or has not been done but in the topological peculiarities of the system. Secondly, the instability we discuss (for magnetics) belongs to the class of dissipative ones. As for the turbulence examples, there (this is obvious from the way the problem is formulated) the generation of large-scale fields is studied in a medium with small-scale helical turbulence. Dissipation may seem unnecessary in the case of magnetics since the inhomogeneous component U_{inh} of the potential energy already contains a 'helical' term $\boldsymbol{M}\,\mathrm{rot}\,\boldsymbol{M}$ (where $\boldsymbol{M}$ denotes the magnetic moment). However, although the final state can be predicted even with dissipation ignored, the evolution scenario cannot be understood if there are no non-conservative terms in the Lifshits–Landau-type equation. From the physics point of view, the need for dissipation is obvious. Indeed, the 'helical terms' provide merging of small-scale subsystems but this process is inelastic (compare with inelastic collision) and can be correctly described only if dissipation is taken into account. In view of these two properties, we use for the mechanism of generation of such topologically non-trivial structures the term *helical-dissipative*. It is necessary to emphasize again that in the case of magnetics, a laminary process is meant.

Let us describe the properties and generation of helicoidal magnetic structures. Note first that according to the results of the analysis given in the classical monograph by Landau and Lifshits (1984), the final magnetic structure must be helicoidal and must minimize the 'energy of inhomogeneity' of the ferromagnetic substance, which corresponds to a rotation of the magnetic moment. This analysis assumed that the value of the magnetic moment is dictated by magnetic forces, which in this case are the main ones. However, on approaching the Curie point, and even more so after passing through it, this situation changes. If only exchange forces were taken into account, the ferromagnetic substance would then transform to a chaotic state. However, if the helicoidal term is included (this term is possible for the crystalline T class), we can expect spontaneous generation of a large-scale helicoidal magnetic structure above the Curie point; in addition to its principal significance as a physical phenomenon, this effect would undoubtedly find practical applications.

Let us now consider the process of build-up of the helicoidal magnetic structure (we follow Merzliakov and Moiseev 1987). Evolution of spin waves with non-zero dissipation is described by the equation (see, for example, Akhiezer *et al* 1968)

$$\frac{\partial \boldsymbol{M}}{\partial t} = \gamma[\boldsymbol{M} \times \boldsymbol{H}_{ef}] + \frac{1}{\tau}\boldsymbol{H}_{ef} - \frac{1}{\tau}[\boldsymbol{n} \times [\boldsymbol{n} \times \boldsymbol{H}_{ef}]]. \tag{6.3.1}$$

Here $\boldsymbol{M}$ is the magnetic moment, γ is the gyromagnetic ratio, $\boldsymbol{n} = \boldsymbol{M}/M$, only one characteristic relaxation time τ is used, and the effective field $\boldsymbol{H}_{ef}$ is defined by

$$\boldsymbol{H}_{ef} = -2\gamma_1 \operatorname{rot} \boldsymbol{M} + \alpha \Delta \boldsymbol{M}$$

where the coefficients γ_1 and α are determined by the contributions to the 'inhomogeneity energy' by the helicoidal and gradient terms, respectively. Assuming $\boldsymbol{M} = \boldsymbol{M}_0 + \boldsymbol{m}$, $\boldsymbol{n}_0 = (0, 0, 1)$ and linearizing equation (6.3.1) with respect to $\boldsymbol{m}$ chosen as a planar wave ($\boldsymbol{m} \sim \exp\{i(\boldsymbol{kr} - \omega t)\}$), where for the sake of simplicity we assume $\boldsymbol{k}_\perp = 0$, we obtain the following dispersion equation:

$$\omega - 2\gamma\gamma_1 M_0 k = \pm\gamma M_0 \alpha k^2 - i\left(\mp\frac{2\gamma_1 k}{\tau} - \frac{\alpha k^2}{\tau}\right). \tag{6.3.2}$$

Equation (6.3.2) shows that instability ($\operatorname{Im}\omega > 0$) occurs if $\alpha k^2 < 2\gamma_1 k$. Therefore, by taking into account dissipative effects it becomes possible to reveal the instability of the homogeneous state of magnetization if the free energy of expression for a magnetic crystal contains terms that correspond to the spin–orbital interaction ($\gamma_1 \neq 0$). It is also of interest that the tendency to reach the minimum of potential energy appears only if dissipation is included in the model. We believe that this is a profound observation that deserves further analysis and bears thinking about in later publications.

It is also important to look closely at the following unusual feature: even though generation and evolution of structures proceed at maximum speed owing to dissipative forces while the equilibrium state is sustained by conservative forces (they determine the minimum of potential energy), only perturbation is responsible for the maximum rate of evolution of structures and also the minimum of potential energy (this is shown in Landau and Lifshits 1982 and follows from (6.3.2)). The following question thus arises: does there exist a universal analytical expression which in the limiting cases of structure formation describes both the initial stage of generation and the saturation stage? What is involved is the creation of an analog to the quantitative theory of irreversible processes for the important case of evolution under destabilization conditions.

2. Sufficiently general conditions of helical vortex dynamo can be formulated on the basis of the results given in the book: (a) a non-zero mean helicity, (b) inhomogeneous parameters of incompressible hydrodynamic

medium, or the medium being two-component, or its compressibility being taken into account. The formation of a helical magnetic-vortex dynamo must be researched later. It is also necessary to analyze the above-mentioned dynamo types not in the two-scale approximation but by a consistent analysis of mode interaction in the conditions of inverse cascade.

3. In section 6.2 we discussed the possible decrease in turbulent viscosity and were able to show that this effect may be observable. However, we do not know whether negative viscosity is possible in media with helical turbulence.

4. Note that contrary to the prevailing opinion, vortices in a helical medium under strong turbulence conditions 'perish' after only one revolution and long-lived vortices appear (well defined quasi-particles)[1] as a consequence of a reduction in turbulent Reynolds stress of the type of $\langle V_x^T V_y^T \rangle$. In this connection it becomes important to attack the problem of consistent construction of perturbation theory for an analysis of strong turbulence.

5. It would be interesting to develop methods of analysis for the multiplicative interaction of modes under conditions of inverse transfer of turbulent energy.

6. Despite the success of the analytical approach to the problems, the necessity to carry out direct numerical simulation of helical vortex dynamics appears compelling both to confirm earlier results and to search for new effects.

References

Akhiezer A I, Bar'yakhtar V G and Peletlinskii S V 1968 *Spin Waves* (Amsterdam: North-Holland)

Batchelor G K 1969 *Phys. Fluids* **12** II–233–9

Belian A V, Moiseev S S and Chkhetiani O G 1994 *Dokl. Akad. Nauk* **334** (1) 34–6

Belian A V, Moiseev S S, Pungin V G and Chkhetiani O G 1997 *Preprint No 1961* (Moscow: Space Research Institute) pp 1–8 (in Russian)

Chamberlian J W 1978 *Theory of Planetary Atmospheres* (New York: Academic)

Dzyaloshinskii I E 1964 *Zh. Exp. Teor. Fiz.* **46** (4) 1420

Fortov V E, Gnedin Yu N, Ivanov M F, Ivlev A V and Klumov B A 1996 *Physics–Uspekhi* **39** (4) 363–92

Hammel H B *et al* 1995 *Science* **267** 1288–97

Henoch C, Hoffert M, Baron A, Klaiman D, Sukoriansky S and Branover H 1990 *Heat Transfer* **4** 9–13

Ingersoll A P and Kanamori H 1995 *Nature* **374** 706–8

Ivanov M F, Galburt V A and Fortov V E 1996 *JETP Lett.* **63** 813–7

Kamenets F F, Petviashvili V I and Pukhov A M 1993 *Izv. RAN FAO* **29** 457–63

Kamenets F F, Pukhov A M, Ivanov M F and Fortov V E 1994 *JETP Lett.* **60** 393

Kamenets F F, Korobov I I and Onischenko O G 1996 *JETP Lett.* **64** 350–356

[1] This problem is currently being worked on by G Branover, B Goldbaum, S Moiseev and A Eidelman.

Kamenets F F, Korobov I I, Ivanov M F, Fortov V E, Galburt V A, Moiseev S S and Onishchenko O G 1997 *Electromagn. Waves and Electr. Systems* **2** (5) 3–12

Keller B S and Yarovskaya I M 1976 *Aerodynamics and Gas Dynamics* (Moscow: Nauka) pp 256–79

Klumov B A, Kondaurov V A, Konyukhov A V, Medvedev Yu D, Sokolovskii A G, Utyuzhnikov S V and Fortov V E 1994 *Physics–Uspekhi* **37** (6) 577–88

Klyatskin V I 1975 *Statisticheskoe Opisanie Dinamicheskih Sistem s Fluktuiruyuschimi Parametrami* (Moscow: Nauka) pp 1–239 (in Russian)

Kraichnan R H 1967 *Phys. Fluids* **10** 1417–23

Landau L D and Lifshits E M 1984 *Electrodynamics of Continuous Media* (Oxford: Pergamon)

Leith C E 1968 *Phys. Fluids* **11** 671–3

Lilly D K 1969 *Phys. Fluids* **12** II–240–9

Merzliakov E G and Moiseev S S 1987 *Preprint No 1155* (Moscow: Space Research Institute) pp 1–19 (in Russian)

Marov M Ya and Kolesnichenko A V 1987 *Introduction to the Planetary Aeronomics* (Moscow: Nauka)

Moiseev S S 1990 *Sov. J. Plasma Phys.* **16** 553–60

Moiseev S S and Chkhetiani O G 1996 *JETP* **83** (1) 192–8

Moiseev S S and Pungin V G 1996 *JETP* **83** (1) 87–94

Moiseev S S, Sagdeev R Z, Tur A V, Khomenko G A and Shukurov A M 1983 *Sov. Phys.–Dokl.* **28** 926–30

Monin A S 1956 *Izv. Akad. Nauk SSSR Ser. Geofiz.* No 4 240–9

Nezlin M V and Snezhkin E N 1993 *Rossby Vortices, Spiral Structures, Solitons: Astrophysics and Plasma Physics in Shallow Water Experiments* (Berlin: Springer)

Ozmidov R V, Belyaev V S and Yampolsky A D 1970 *Izv. Akad. Nauk SSSR FAO* **6** (3) 285–91

Pedlosky J 1987 *Geophysical Fluid Dynamics* (New York: Springer)

Starr V 1953 *Tellus* **5** 494–8

Starr V 1968 *Physics of Negative Viscosity Phenomena* (New York: McGraw–Hill)

Vainshtein S I, Zeldovich Ya B and Ruzmaikin A A 1980 *Turbulentnoe Dynamo v Astrofizike* (Moscow: Nauka) pp 1–352 (in Russian)

Webster F 1965 *Tellus* **17** 239–45

Index

T - #0197 - 101024 - C0 - 229/152/9 [11] - CB - 9780750304832 - Gloss Lamination